PROTEIN STABILITY AND STABILIZATION THROUGH PROTEIN ENGINEERING

ELLIS HORWOOD SERIES IN BIOCHEMISTRY AND BIOTECHNOLOGY

Series Editor: Dr ALAN WISEMAN, Senior Lecturer in Biochemistry, University of Surrey, Guildford

Aitken, A.	Identification of Protein Consensus Sequences
Ambrose, E.J.	The Nature and Origin of the Biological World
Austin, B. & Brown, C.M.	Microbial Biotechnology: Freshwater and Marine Environments*
Berkeley, R.C.W. *et al.*	Microbial Adhesion to Surfaces
Bertoncello, I.	Human Cell Cultures for Screening Anti-Cancer Rays*
Blackburn, F. & Knapp, J.S.	Agricultural Microbiology*
Bowen, W.R.	Membrane Separation Processes*
Bubel, A. & Fitzsimons, C.	Microstructure and Function of Cells
Clark, C.R. & Moos, W. H.	Drug Discovery Technologies
Cook, N.	Potassium Channels
Corkill, J.A.	Clinical Biochemistry: The Analysis of Biologically Important Compounds and Drugs*
Dolly, J.O.	Neurotoxins in Neurochemistry
Espinal, J.	Understanding Insulin Action: Principles and Molecular Mechanisms
Eyzaguirre, J.	Chemical Modification of Enzymes
Eyzaguirre, J.	Human Biochemistry and Molecular Biology
Ferencik, M.	Immunochemistry*
Firman, K.	DNA Cloning/Sequencing Workshop: A Short Course
Fish, N.M.	Computer Applications in Fermentation Technology*
Francis, J.L.	Haemostasis and Cancer*
Gacesa, P. & Russell, N.J.	Pseudomonas Infection and Alginates: Structures, Properties and Functions in Pathogenesis*
Gemeiner, P. *et al.*	Enzyme Engineering: Immobilized Biosystems
Ghose, T.K.	Bioprocess Computations in Biotechnology: Vol. 1
Ghose, T.K.	Bioprocess Engineering: The First Generation
Griffin, R.L.	Using the Transmission Electron Microscope in the Biological Sciences
Harding, J.	Biochemistry and Pharmacology of Cataract Research: Drug Therapy Approaches*
Horobin, R.W.	Understanding Histochemistry: Selection, Evaluation and Design of Biological Stains
Hudson, M.J. & Pyle, P.L.	Separations for Biotechnology, Vol 2*
Junter, G.A.	Electrochemical Detection Techniques in the Applied Biosciences: Vol. 1: Analysis and Clinical Applications Vol. 2: Fermentation and Bioprocess Control, Hygiene and Environmental Sciences
Kennedy, J.F. & White, C.A.	Bioactive Carbohydrates
Krcmery, V.	Antibiotic and Chemotherapeutic Compounds*
Krstulovic, A. M.	Chiral Separations by HPLC
Lembeck, F.	Scientific Alternatives to Animal Experiments
Nosoh, Y. & Sekiguchi, T.	Protein Stability and Stabilization through Protein Engineering
Palmer, T.	Understanding Enzymes, Third Edition
Reizer, J. & Peterkofsky, A.	Sugar Transport and Metabolism in Gram-Positive Bacteria
Russell, N.J.	Microbes and Temperature
Scragg, A.H.	Biotechnology for Engineers: Biological Systems in Technological Processes
Scragg, A.H.	Bioreactors in Biotechnology: A Practical Approach
Sikyta, B.	Methods in Industrial Microbiology
Sluyser, M.	Asbestos-Related Cancer
Sluyser, M.	Molecular Biology of Cancer Genes
Sluyser, M. & Voûte, P.A.	Molecular Biology and Genetics of Childhood Cancers: Approaches to Neuroblastoma
Verrall, M.S.	Discovery and Isolation of Microbial Products
Verrall, M.S. & Hudson, M.J.	Separations for Biotechnology
Walum, E., Stenberg, K. & Jenssen, D.	Understanding Cell Toxicology: Principles and Practice
Webb, C. & Mavituna, F.	Plant and Animal Cells: Process Possibilities
Winkler, M.	Biochemical Process Engineering*
Wiseman, A.	Handbook of Enzyme Biotechnology, 2nd Edition
Wiseman, A.	Topics in Enzyme and Fermentation Biotechnology Vols 1–10
Wiseman, A.	Enzyme Induction, Mutagen Activation and Carcinogen Testing in Yeast
Wiseman, A.	Genetically Engineered Proteins and Enzymes from Yeasts: Production Control

* *In preparation*

PROTEIN STABILITY AND STABILIZATION THROUGH PROTEIN ENGINEERING

YOSHIAKI NOSOH Ph.D.
TAKESHI SEKIGUCHI Ph.D.
Department of Fundamental Science
College of Science and Engineering
Iwaki Meisei University, Fukushima, Japan

ELLIS HORWOOD
NEW YORK LONDON TORONTO SYDNEY TOKYO SINGAPORE

First published in 1991 by
ELLIS HORWOOD LIMITED
Market Cross House, Cooper Street,
Chichester, West Sussex, PO19 1EB, England

A division of
Simon & Schuster International Group
A Paramount Communications Company

© Ellis Horwood Limited, 1991

All rights reserved. No part of this publication may be reproduced, stored in a retrieval system, or transmitted, in any form, or by any means, electronic, mechanical, photocopying, recording or otherwise, without the prior permission, in writing, of the publisher

Every effort has been made to trace all copyright holders, but if any have been inadvertently overlooked, the publishers will be pleased to make the necessary arrangements at the earliest opportunity.

Printed and bound in Great Britain
by Bookcraft Ltd, Midsomer Norton, Avon

British Library Cataloguing-in-Publication Data

Nosoh, Yoshiaki
Protein stability and stabilization through protein engineering. —
(Ellis Horwood series in biochemistry and biotechnology)
I. Title. II. Sekiguchi, Takeshi. III. Series.
574.19
ISBN 0–13–721788–9

Library of Congress Cataloging-in-Publication Data available

Table of contents

1	Introduction		9
	References		11
2	Protein structure		15
	2.1	Amino acids	15
	2.2	Polypeptide chain	16
		2.2.1 Peptide formation	16
		2.2.2 Peptide structure	20
		2.2.3 Constraint of dihedral angles	23
	2.3	Secondary structure	26
		2.3.1 Factors determining protein structure	26
		2.3.1.1 Dispersion forces	27
		2.3.1.2 Electrostatic interactions	27
		2.3.1.3 van der Waals potentials	27
		2.3.1.4 Hydrogen bonds	28
		2.3.1.5 Hydrophobic interactions	29
		2.3.2 Helices	31
		2.3.3 β-Sheets	34
		2.3.4 Non-repetitive structures	38
		2.3.5 Supersecondary structures	40
	2.4	Tertiary structure	41
		2.4.1 Structural domains	41
		2.4.2 Structural types	42
		2.4.3 Roles of individual amino acids	46
	2.5	Quaternary structure	57
	2.6	Side chains and water	60
	2.7	Prediction of protein structures	62
		2.7.1 Secondary structure	62

		2.7.2	Tertiary structure	65
	References			66
3	Protein stability			79
	3.1	Thermodynamics of protein unfolding		79
		3.1.1	Thermodynamic parameters	80
		3.1.2	Determination of ΔG	81
	3.2	Stabilizing forces		82
		3.2.1	Dispersion forces	83
		3.2.2	Conformational entropy	83
		3.2.3	Hydrophobic interactions	85
		3.2.4	Hydrogen bonding	86
		3.2.5	Electrostatic interactions	87
		3.2.6	Structural features	88
	References			90
4	Stable proteins			101
	4.1	Use of stable proteins		101
	4.2	Thermophilic bacteria		102
	4.3	Thermophilic proteins		102
	4.4	Mechanism of stability of thermophilic proteins		103
		4.4.1	Replacement of many amino acids	104
			4.4.1.1 Hydrophobic interactions	104
			4.4.1.2 Compact packing or rigidity	109
			4.4.1.3 Electrostatic interactions	110
			4.4.1.4 Hydrogen bonds	110
		4.4.2	Substitution of few amino acids	111
			4.4.2.1 Disulfide bonds	112
			4.4.2.2 Electrostatic interactions	112
			4.4.2.3 Hydrophobic interactions	112
			4.4.2.4 Hydrogen bonds	113
	References			114
5	Stabilization through chemical and physical modifications			124
	5.1	Strategies for protein stabilization		125
		5.1.1	Chemical modification	125
		5.1.2	Immobilization	126
	5.2	Evaluation of three strategies for stabilizing proteins		127
	5.3	Stabilization by chemical modification		128
		5.3.1	Possible reasons for altered thermostability	128
		5.3.2	Basis of thermostability in guanidinated proteins	130
			5.3.2.1 Subunit–subunit interaction	130
			5.3.2.2 Compact packing	132
	References			134
6	Stabilization through biological modification			140
	6.1	Site-directed mutagenesis		141
	6.2	Mutagenesis with knowledge of protein structure		143
		6.2.1	Altering destabilizing factors	143
			6.2.1.1 Introduction of disulfide bonds	144
			6.2.1.2 Other amino acid substitutions	157

	6.2.2	Altering stabilizing factors	161
		6.2.2.1 Internal hydrophobicity	161
		6.2.2.2 Electrostatic interactions	165
		6.3.2.3 Hydrogen bonds	167
	6.2.3	Compensating stabilizing–destabilizing factors	170
	6.2.4	Protection against chemical destabilization	172
6.3	Mutagenesis without knowledge of protein structures		176
	6.3.1	Use of a homologous model protein	176
	6.3.2	Mutagenesis without a homologous protein	180
6.4	Stabilization through other mutagenesis		184
	6.4.1	Chemical mutagenesis	184
	6.4.2	Thermal mutagenesis	184
References			186
7 Concluding remarks			197
References			209
Index			219

1

Introduction

Proteins, nucleic acids, polysaccharides and lipid assemblies are the four major biopolymers (Lehninger, 1982). Because of a variety of important biological functions (Darnell et al., 1986), proteins and nucleic acids have attracted much more attention from biological scientists than have polysaccharides and lipid assemblies (membranes). Innumerable numbers of books, reviews and papers have been published on the structure and function of proteins (e.g., Creighton, 1983; Sanger, 1983; Blake and Johnson, 1984; Fersht, 1985; Tonegawa, 1985) and of nucleic acids, especially DNA (e.g. Elsenfeld, 1985). This does not mean that the functions of polysaccharides and membranes, especially the latter, are physiologically less important than those of proteins and nucleic acids. In fact, membranes play various important functions essential for living activity of procaryotic and eucaryotic organisms (e.g. Harison and Lunt, 1980; Finean et al., 1984). The physiologically important functions of membranes, e.g. transportation (Wilson, 1978), cell-to-cell signaling (Berridge, 1985) and energy conversion (Ferguson and Sorgato, 1982), mostly depend on the proteins which are immobilized or attached to membranes.

In spite of a great variety of protein functions, e.g. as a catalyst for biochemical reactions (Dixon and Webb, 1979), as a receptor to hormone proteins (e.g. Gilman, 1984) and as an antibody in immune systems (e.g. Davies and Metzger, 1983), and divergent specificities of each function (e.g. substrate specificity of enzymes), naturally occurring proteins all consist of only 20 kinds of α- and L-amino acids. This gives rise to an important question; i.e. how proteins having a great number of functions and specificities can be constructed from only the 20 amino acids, and how proteins of three-dimensional structures can correspond to their peculiar function and specificity. It is now understood that a specific function of proteins is reflected by its peculiar three-dimensional structure and that the peculiar tertiary structure is determined by its amino acid sequence (Schulz and Schirmer 1979; Dickerson and Geis, 1980). The evidence from molecular biology shows that the nucleotide sequence on the gene DNA coding for a protein is first transcribed to mRNA, and then the transcribed genetic information on mRNA is translated into the amino acid sequence of the protein, through a complex protein synthesis machinery involving ribosomes (Lehninger, 1982). Through the translation process, the amino acids are sequentially chemically combined, i.e. peptide bonded, according to the genetic code in mRNA (Haselkorn and Rothman-Denes, 1973).

In the translation process, natural substitution of even a single nucleotide on the gene DNA (mutation), and thus a single nucleotide substitution of mRNA, produces a mutant protein in which a single amino acid is substituted (Old and Primrose, 1981; Watson et al., 1987). If the amino acid substitution occurs at the site affecting protein function, the original function of the protein is modified (e.g. Gorini et al., 1961). If the site is related to the thermostability of the protein, the stability is reduced, producing temperature-sensitive mutants (Reed et al., 1985). This is the cause of genetic diseases due to loss of function and the reason for the appearance of numerous numbers of mutants of microorganisms with modified function and stability (Watson et al., 1987).

Another question is how proteins can be stabilized. Elaborate and delicate conformations of proteins must be kept stable while they are functioning in physiological environments. Therefore qualitative and quantitative approaches to protein stability have long been made with model organic compounds, oligopeptides and proteins *per se* or their mutant proteins of different stability (Schulz and Schirmer, 1979; Creighton, 1983). The X-ray crystallographic analysis (Sherwood, 1976; Blundell and Johnson, 1976) of proteins, of course, had greatly assisted these approaches (Schulz and Schirmer, 1979; Creighton, 1983; Alber, 1989). It is now accepted that the forces or interactions which stabilize proteins are van der Waals contacts, hydrophobic contacts, hydrogen bonds and electrostatic interactions between polypeptide chains, between peptide chain and amino acid residue, and between amino acid residues, Proteins are stabilized as a result of the sum of the forces (Schulz and Schirmer, 1979; Creighton, 1983; Alber, 1989).

Proteins from thermophiles are more stable than the corresponding mesophilic proteins (e.g. Zuber, 1976). Analyses of enhanced stability of thermophilic proteins may provide some additional information on protein stability. However, additional stability of thermophilic proteins relative to mesophilic ones seems to be caused differently from protein to protein (Menedez-Arias and Argos, 1989; Nosoh and Sekiguchi, 1988, 1990).

Recent developments in recombinant DNA technique (Abelson and Butz, 1990) have made it possible to substitute any amino acid in a protein at will, through artificial replacement of a nucleotide in the gene encoding the protein (Jackson et al., 1972; Grobstein, 1977). The procedure is often called site-directed mutagenesis. The strategy of modifying the function of a protein through site-directed mutagenesis is called protein engineering (Ulmer, 1983). Through protein engineering various properties of proteins such as specific activity and substrate specificity of enzymes can be modified. This strategy is now becoming a promising tool for revealing the molecular mechanisms of various functions of proteins, and for improving the functions of industrial and medical proteins, when the three-dimensional structure of a protein is known and information on its structure–function relationship is available (Hanna, 1986; Holmes, 1986). Similarly, protein stability can be modified by site-directed mutagenesis (Nosoh and Sekiguchi, 1988, 1990; Alber, 1989). In addition, if amino acid substitution does not affect protein conformation, one can specify and influence the interaction(s) around the substitution site to change and thus dissect protein stability.

There have been few books relating the two topics of protein stability and stabilization through protein engineering (Jardetzki, 1989). The two topics are indeed the two sides of a coin. In this book then we have tried to pay much attention to relating protein stability with protein stabilization.

First, the architecture of proteins, which is fundamental for an understanding of protein stability and stabilization, is described (Chapter 2). The X-ray crystal analysis of proteins has greatly contributed to establishing their three-dimensional structure. Since information on protein structures revealed by highly refined X-ray crystallographic determination has been accumulated, therefore, this chapter deals with recent information on protein structures as well as that obtained previously.

In Chapter 3, protein stability is discussed qualitatively and quantitatively. Thermodynamic treatment of protein unfolding is indeed important for quantitatively estimating protein stability. The thermodynamics of protein unfolding is then described. Proteins are principally stabilized by various physical factors, contacts or interactions, which should be thermodynamically or quantitatively evaluated. The results so far obtained with this approach and the results recently accumulated through protein engineering are also included here.

In Chapter 4 is described the necessity for stable enzymes, especially with respect to thermal denaturation or inactivation, from fundamental and applied points of view—for analyzing protein stability and for biotechnological uses, respectively. Thus thermophiles as sources of stable enzymes are briefly introduced, together with the mechanisms of so far presented for the thermostability of thermophilic proteins.

Chapter 5 is concerned with the three principal strategies for stabilizing proteins, together with evaluation of the procedures. Chemical modification of amino acid residues in proteins has long been employed to modify the functions, including stability, of proteins. The stabilization and stability of proteins through chemical modification are then described. Nowadays, however, the most promising strategy for analyzing protein stability and for stabilizing proteins is amino acid substitution by site-directed mutagenesis (protein engineering).

Stabilization of proteins, by which the mechanism of protein stability is revealed, is introduced and detailed results given in Chapter 6.

The book closes by summarizing and criticizing the preceding chapters, and the future aspects of protein engineering in the fundamental and engineering fields are then discussed. In writing Chapters 2 and 3 the work of Schulz and Schirmer (1979) and Fasman (1989) were referred to.

REFERENCES

Abelson, J. and Butz, E. (eds) (1980) Recombinant DNA. *Science*, **209**, 1317–1338.

Alber, T. (1989) Stabilization energies of protein conformation. In *Prediction of protein structure and the principles of protein stabilization* (G. D. Fasman, ed.), pp. 161–192. Plenum, New York.

Berridge, M. (1985) The molecular basis of communication within the cell. *Sci. Am.*, **253**, 142–150.

Blake, C. C. F. and Johnson, L. N. (1984) Protein structure. *Trends Biochem. Sci.*, **9**, 147–151.

Blundell, T. L. and Johnson, L. N. (1976) *Protein crystallography.* Academic Press, New York.

Creighton, T. E. (1983) *Proteins.* Freeman, San Francisco.

Darnell, J., Lodish, H. and Baltimore, D. (1986) *Molecular cell biology.* Freeman, New York.

Davies, D. R. and Metzger, H. (1983) Structural basis of antibody function. *Annu. Rev. Immunol.*, **1**, 87–117.

Dickerson, R. E. and Geis, I. (1980) *Proteins: structure, function, and evolution.* Benjamin/Cummings, Menlo Park.

Dixon, M. and Webb, E. C. (1979) *Enzymes*, 3rd. edn. Longman, London.

Elsenfeld, G. (1985) DNA. *Sci. Am.*, **253**, 58–66.

Fasman, G. D. (ed.) (1989) *Prediction of protein structure and the principles of protein conformation.* Plenum, New York.

Ferguson, S. J. and Sorgato, M. C. (1982) Proton electrochemical gradients and energy transduction processes. *Annu. Rev. Biochem.*, **51**, 185–218.

Fersht, A. (1985) *Enzyme structure and metabolism*, 2nd. edn. Freeman, San Francisco.

Finean, J. B., Coleman, R. and Michell, R. H. (1984) *Membranes and their cellular function*, 3rd. edn. Blackwell, New York.

Gilman, A. G. (1984) G proteins and dual control of adenylate cyclase. *Cell*, **36**, 571–579.

Gorini, L., Gundersen, W. and Burger, M. (1961) Genetics of regulation of enzyme structure synthesis in arginine biosynthetic pathway of *E. coli. Cold Spring Harbor Symp. Quant. Biol.*, **26**, 173–182.

Grobstein, C. (1977) The recombinant-DNA debate. *Sci. Am.*, **237**, 22–33.

Hanna, M. H. (1986) Applied genetics for biochemical engineering: recombinant DNA. In *Advanced biochemical engineering* (H. R. Bungay and Belfort, G., eds), pp. 103–128. Wiley, New York.

Harrison, R. and Lunt, G. G. (1980) *Biological membranes, their structure and function*, 2nd edn. Halsted, New York.

Haselkorn, R. and Rothman-Denes, L. B. (1973) Protein synthesis. *Annu. Rev. Biochem.*, **42**, 397–438.

Holmes, D. S. (1986) Molecular enzyme engineering. In *Advanced biochemical engineering* (H. R. Bungay and Belfort, G., eds), pp. 129–165. Wiley, New York.

Jackson, D., Symons, R. and Berg, P. (1972) Biochemical method for inserting new genetic information into DNA of simian virus 40: circular SV40 DNA molecules containing lambda phage genes and the galactose operon of E. coli. *Proc. Natl. Acad. Sci. USA*, **69**, 2904–2909.

Jardetzki, O. (ed.) (1989) *Protein structure and engineering*. Plenum, New York.

Karp, G. (1984) *Cell biology*, 2nd. edn. McGraw-Hill, New York.

Lehninger, A. L. (1982) *Principles of biochemistry*. Worth, New York.

Menedez-Arias, L. and Argos, P. (1989) Engineering protein thermal stability. *J. Mol. Biol.*, **206**, 397–406.

Nosoh, Y. and Sekiguchi, T. (1988) Protein thermostability: mechanism and control through protein engineering. *Biocatalysis*, **1**, 257–273.

Nosoh, Y. and Sekiguchi, T. (1990) Protein engineering for thermostability. *Trends Biotech.*, **8**, 16–20.

Old, R. W. and Primrose, S. B. (1981) *Principles of gene manipulation*, 2nd edn. Blackwell, London.

Reed, J. A., Hadwiger, J. A. and Lorincz, A. T. (1985) Protein kinase activity associated with the product of the yeast cell division cycle gene *cdc28*. *Proc. Natl. Acad. Sci. USA*, **88**, 4055–4059.

Sanger, W. (1983) *Principles of nucleic acid structure*. Springer-Verlag, New York.

Schulz, G. E. and Schirmer, R. H. (1979) *Principles of protein structure*. Springer-Verlag, New York.

Sherwood, D. (1976) *Crystals, x-rays and proteins*. Wiley, New York.

Tonegawa, W. (1985) The molecules of the immune system. *Sci. Am.*, **253**, 122–130.

Ulmer, K. M. (1983) Protein engineering. *Science*, **219**, 666–676.

Watson, J. D., Hopkins, N. H., Roberts, J. W., Steitz, J. A. and Weiner, A. M. (1987) *Molecular biology of the gene*. Benjamin/Cummings, Menlo Park.

Wilson, D. B. (1978) Cellular transport mechanism. *Annu. Rev. Biochem.*, **47**, 933–965.

Zuber, H. (ed.) (1976) *Enzymes and proteins from thermophilic microorganisms*. Birkhauser Verlag, Basel.

2
Protein structure

Numerous proteins have so far been isolated and purified from various kinds of living organisms. For examples, several thousands of enzymes—proteins exhibiting catalytic activity—have been purified to a homogeneous state and the amino acid compositions of almost all of them have been determined. In spite of the enormous number of naturally occurring proteins, the kinds of amino acids constructing the proteins number only 20. The vast difference in three-dimensional structure and, therefore, in function of proteins are originate solely from the different amino acid sequences in proteins; i.e. the peculiar structure of a protein is determined by the chemical and physical properties of the amino acids aligned in the protein sequence (Schulz and Schirmer, 1979; Creighton, 1983).

In this chapter is described the protein conformation necessary for understanding the stability and stabilization of proteins which are described later in this book. For more detailed information on protein structures refer to Schulz and Schirmer (1979) and Creighton (1983).

2.1 AMINO ACIDS

All the amino acids found in naturally occurring proteins are of α-type and L-configuration at C_α-position (Fig. 2.1) (Schulz and Schirmer, 1979; Creighton, 1983). The amino acids of β-type or D-configuration are seldom found in natural proteins.

In Fig. 2.1 is shown the 'CORN crib' for L-amino acids (Richardson and Richardson, 1989). The side chains (R) of all 20 amino acids are listed in Fig. 2.2. They are generally classified into the following categories: hydrophobic or non-polar group consisting of aromatic (Phe, Trp and Pro) and aliphatic side chains (Ala, Val, Leu, Ile and Met); uncharged polar group (Gly, Ser, Thr, Cys, Tyr, Asn and Gln); positively charged polar group (Lys, Arg and His), and negatively charged polar group (Asp and Glu) (Lehninger, 1981). The properties of these residues as they relate to the structure and

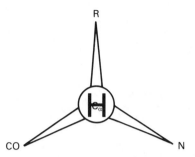

Fig. 2.1—Mnemonic for determining the handedness of an amino acid (*CORN* crib). Looking at the α-carbon from the direction of the H, the other substitutions should read CO (carbonyl), R (side chain), and N (backbone amino) in clockwise order for L-amino acid. (Redrawn from Richardson and Richardson, 1989.)

in contributing to the stability and stabilization of proteins will be discussed in more detail later (section 2.4.2 and Chapter 3).

Among the amino acids Gly and Pro have peculiar conformations, because Gly has no side chain and in Pro an α-amino group is replaced by an imino group (Fig. 2.2). These structural characteristics of the amino acids play unique roles in conformation and stabilization of proteins (see later in this chapter and Chapter 3). The various chemical and physical properties of the amino acids themselves and of those in proteins are reflected by the properties of their side chains (Table 2.1). Therefore, the properties of the side chains play important roles in folding polypeptide chains into three-dimensional (secondary and tertiary) structures (see sections 2.3–2.6 for details).

2.2 POLYPEPTIDE CHAIN

Protein structure can principally be described at four different levels. *Primary structure* refers to the linear arrangement of amino acid residues along a polypeptide chain, and to the locations of covalent bonds, such as disulfide bonds between chains or within a chain. *Secondary structure* pertains to the folding of parts of these chains into regular structures such as α-helices and β-sheets, as described later. Sometimes β-turns are included in this level of protein structure. *Tertiary structure* includes the folding of regions between α-helices and β-sheets, and all of the non-covalent interactions that ensure the proper folding of a single polypeptide chain, such as hydrogen bonds and hydrophobic, electrostatic and van der Waals interactions. Finally, *quaternary structure* refers to the non-covalent interactions that bind several polypeptide chains into a single protein molecule, as seen with many proteins.

2.2.1 Peptide formation

Present knowledge of molecular biology indicates that protein synthesis proceeds as schematically shown in Fig. 2.3.

The nucleotide sequence in a gene encoding a protein is transcribed to messenger RNA (mRNA), and then the nucleotide sequence in mRNA is translated on the ribosome into the amino acid sequence originally encoded by the gene, applying the

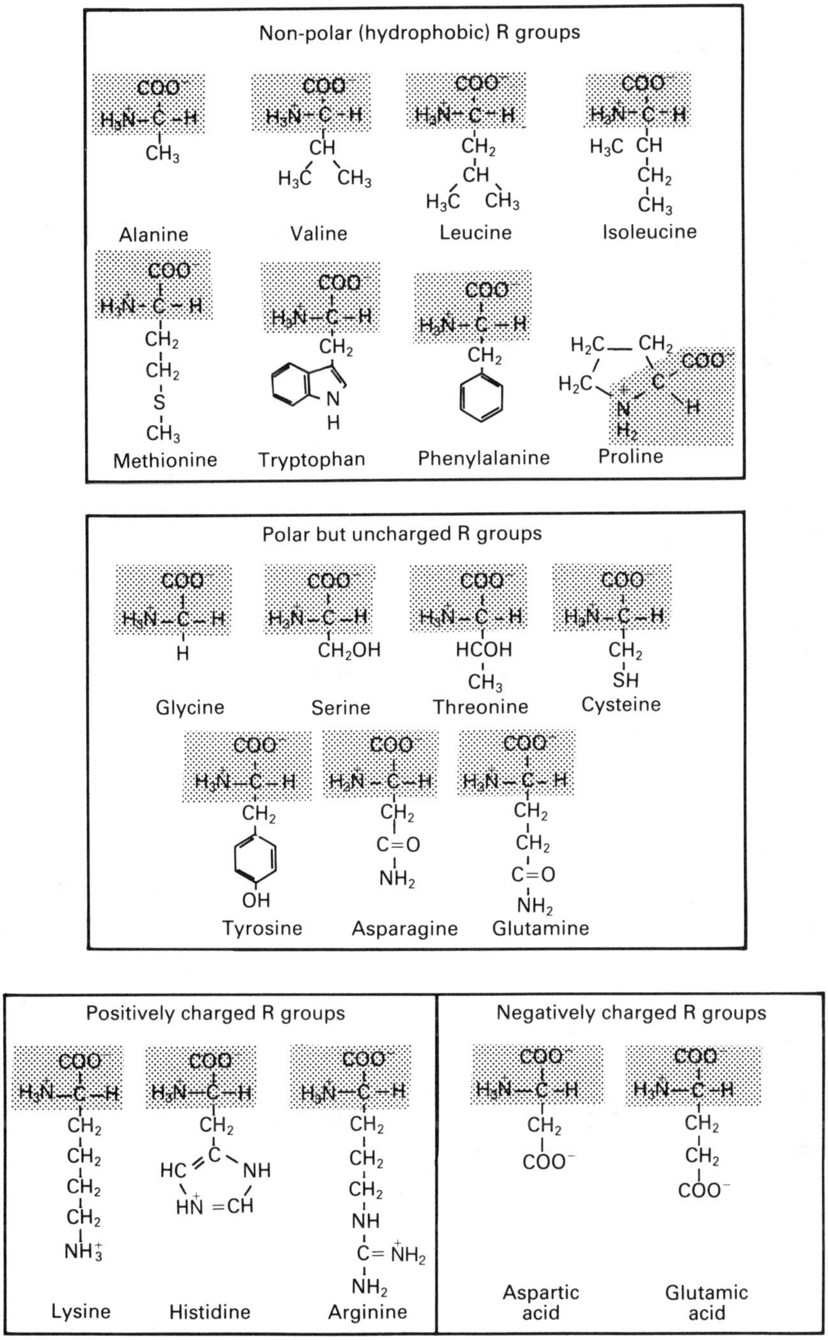

2.2 Polypeptide Chain

Fig. 2.2—The 20 amino acids commonly found in proteins. They are shown with the amino and carboxyl groups ionized, as they occur at pH 7.0. The shaded portions are those common to all the amino acids. The unshaded portions are their R groups.

Table 2.1 — Notations and properties of the standard amino acid residues of proteins[a]

Amino acid	Three-letter symbol	One letter symbol	Mol wt of residue at pH 7.0	pK values for the ionizing groups at 25°C		
				pK_1 —COOH	pK_2 —NH_3^+	pK_R R group
Alanine	Ala	A	71	2.34	9.69	
Glutamate	Glu	E	128	2.19	9.67	4.25
Glutamine	Gln	Q	128	2.17	9.13	
Aspartate	Asp	D	114	2.09	9.82	3.86
Asparagine	Asn	N	114			
Leucine	Leu	L	113	2.36	9.60	
Glycine	Gly	G	57	2.34	9.6	
Lysine	Lys	K	129	2.18	8.95	10.53
Serine	Ser	S	87	2.21	9.15	
Valine	Val	V	99			
Arginine	Arg	R	157	2.17	9.04	12.48
Threonine	Thr	T	101	2.63	10.43	
Proline	Pro	P	97			
Isoleucine	Ile	I	113			
Methionine	Met	M	131			
Phenylalanine	Phe	F	147			
Tyrosine	Tyr	Y	163	2.20	9.11	10.07
Cysteine	Cys	C	103	1.71	10.78	8.33
Tryptophan	Trp	W	186			
Histidine	His	H	137	1.82	9.17	6.0

[a] Asx and Glx are either acid or amide form. B, Z and X denote Asx, Glx and undetermined or non-standard amino acid residue, respectively. R denotes the side chain of the amino acid.

2.2 Polypeptide Chain

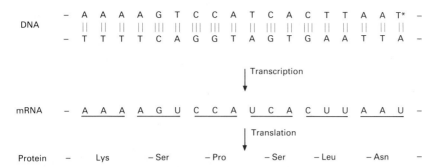

Fig. 2.3—Scheme of protein synthesis. The nucleotide sequence of the gene DNA coding for a protein is transcribed to mRNA, which is then translated to the amino acid sequence of the protein, using the genetic codons shown in Fig. 2.4.

The second letter of the codons

		U		C		A		G	
		UUU	Phe	UCU	Ser	UAU	Tyr	UGU	Cys
	U	UUC	Phe	UCC	Ser	UAC	Tyr	UGC	Cys
		UUA	Leu	UCA	Ser	UAA	End	UGA	End
		UUG	Leu	UCG	Ser	UAG	End	UGG	Trp
The		CUU	Leu	CCU	Pro	CAU	His	CGU	Arg
first		CUC	Leu	CCC	Pro	CAC	His	CGC	Arg
letter	C	CUA	Leu	CCA	Pro	CAA	Gln	CGA	Arg
of the		CUG	Leu	CCG	Pro	CAG	Gln	CGG	Arg
codons		AUU	Ile	ACU	Thr	AAU	Asn	AGU	Ser
(5' end)		AUC	Ile	ACC	Thr	AAC	Asn	AGC	Ser
	A	AUA	Ile	ACA	Thr	AAA	Lys	AGA	Arg
		AUG	Met	ACG	Thr	AAG	Lys	AGG	Arg
		GUU	Val	GCU	Ala	GAU	Asp	GGU	Gly
	G	GUC	Val	GCC	Ala	GAC	Asp	GGC	Gly
		GUA	Val	GCA	Ala	GAA	Glu	GGA	Gly
		GUG	Val	GCG	Ala	GAG	Glu	GGG	Gly

Fig. 2.4—The dictionary of amino acid code words as they occur in mRNAs. The codons are written in the 5' → 3' direction. The third base of each codon is less specific than the first and second bases. The initiation codon AUG is shaded, and the three termination codons, UAA, UAG, and UGA, are enclosed in boxes. All the amino acids except Met and Trp have more than one codon. The codons, as they occur in DNA, will be complementary to the mRNA codons in Fig. 2.4, but will be antiparallel and will contain T residues at positions complementary to A, and A residues at positions complementary to U.

Table 2.2—Components required in protein synthesis in *E. coli* (see Lehninger, 1981 for more detail)

Stage	Necessary components
1. Activation of the amino acids	Amino acids
	tRNAs
	Aminoacyl-tRNA synthetases
	ATP
	Mg^{2+}
2. Initiation of the polypeptide chain	fMet-tRNA in procaryotes
	(Met-tRNA in eucaryotes)
	Initiation codon in mRNA (AUG)
	mRNA
	GTP
	Mg^{2+}
	Initiation factors (IF-1, IF-2, IF-3)
	30S ribosomal subunit
	50S ribosomal subunit
3. Elongation	Functional 70S ribosomes
	Aminoacyl-tRNAs specified by codons
	Mg^{2+}
	Elongation factors (EF-T and EF-G)
	GTP
4. Termination	Functional 70S ribosomes
	Termination codon in mRNA
	Polypeptide release factors (R_1, R_2, R_3)

genetic code (Fig. 2.4) (Lehninger, 1981; Watson *et al.*, 1987). In Table 2.2 are shown the components required in the five major stages in polypeptide synthesis in *Escherichia coli* (Lehninger, 1982). As can be seen from Fig. 2.3, the substitution of one nucleotide in a gene DNA to another, by natural or artificial (protein engineering) mutagenesis, causes the amino acid replacement which results in changes of structure and thus, in some cases, the functional change of the protein (Chapter 6). For example, when the thymine (T) marked as an asterisk on the DNA sequence in Fig. 2.3 is mutated to adenine (A), Asn in the original amino acid sequence is replaced with Lys.

In the translating process the amino acids are successively linked through covalent linkage (peptide bond) (Fig. 2.5). As indicated in the figure, a peptide chain is formed in the direction of the arrow, leaving an amino group of the N-terminal amino acid residue.

2.2.2 Peptide structure

Figure 2.6(a) shows the geometric parameters of the peptide unit and the α-carbon (Momany *et al.*, 1975). The bond lengths and angles shown are the local

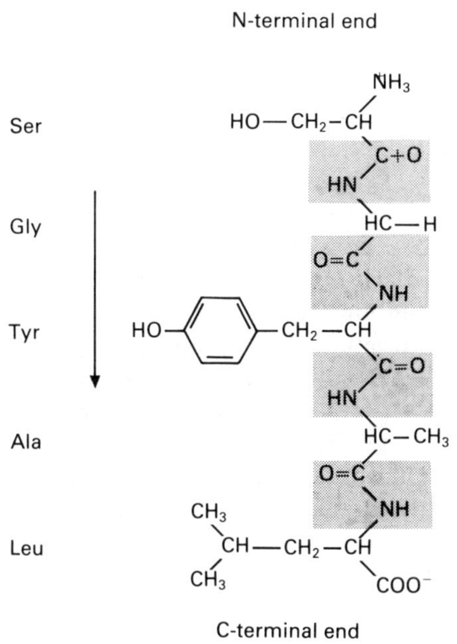

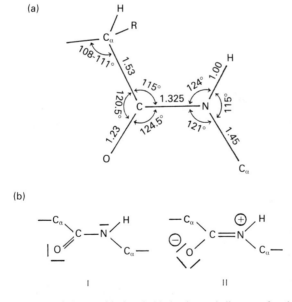

Fig. 2.5—Structure of a polypeptide. In the figure a pentapeptide (Ser–Gly–Tyr–Ala–Leu) is shown. Peptides are synthesized in the direction of the arrow. The peptide bonds are shaded.

Fig. 2.6—The geometry of the peptide bond. (a) Angles and distances for the usual peptide bond as given by Momany et al. (1975). (Redrawn from Richardson and Richardson, 1989.) (b) The two limiting electronic structures of the peptide bond, I allowing free rotation and II restricting rotation but having a large dipole moment. The hybrid structure contains 60% of I and 40% of II (Pauling et al., 1951a).

minimum-energy values around which the structure fluctuates, both as a function of time for a given bond or angle and also statistically among the total set of such bonds or angles.

The distance between the C- and N-atoms of a peptide bond is 0.25 Å or 10% shorter than the normal C—N bond, and the C=O double bond is 0.02 Å longer than those of aldehydes and ketones (Pauling et al., 1951). These results are explained by the resonance between the two structures shown in Fig. 2.6(b). In structure I the C—N bond contains only axial symmetric σ-electrons allowing free rotation, while structure II has σ- and π-electrons in the C—N bond, giving rise to a large dipole moment and inhibiting rotation. The π-electrons are smeared out over the C—O and C—N bonds. Since structure II is planar, the hybrid structure is also planar and the six atoms, C_α, C, O, N, C_α and H, lie in one plane (Fig. 2.6a).

The conformation of a polypeptide chain is described by the three backbone dihedral angles per residue (ϕ, ψ and ω) and the dihedral angle χ from the side chain (Fig. 2.7a). In Fig. 2.7(b) is illustrated the standard convention for measuring dihedral angles (IUPAC, 1970). A dihedral angle involves four successive atoms—A, B, C and D—and the three bonds joining them. If one looks directly down the length of the central bond joining atoms B and C, with atom A nearest at 12 on the clock face,

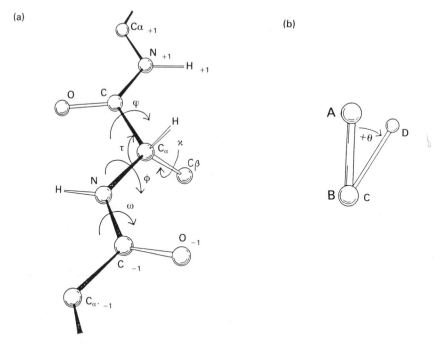

Fig. 2.7—Nomenclature of the atoms of the polypeptide chain, the tetrahedral bond angle τ, and the backbone dihedral angles $\phi, \psi,$ and ω (a). (b) Illustration of the standard convention (IUPAC, 1970) for measuring dihedral angles. The dihedral, or torsion, angle around a bond B—C is defined by the relative positions of the four atoms A, B, C, and D. Looking down the B—C bond, atom A is placed at 12 o'clock, and atom D measures the dihedral angle: plus if clockwise, as in this example (about $+35°$), and minus if counterclockwise. (Redrawn from Richardson and Richardson, 1989.)

the clock position of atom D gives the angle. By convention, dihedral angles are assigned in the range $-180°$ to $+180°$, with the clockwise direction being positive. The dihedral angle shown in Fig. 2.7(b) is about $+35°$.

The conformational description of peptides as shown in Fig. 2.7 is ideal for comparing short pieces of structure, but it is not applicable for specifying global conformation. Even very small round-off errors accumulate drastically. As mentioned above, a peptide bond is planar so that ω (Fig. 2.7a) is always within about $10°$ of $180°$, which is fully extended or in *trans* conformation. The curled-up *cis* conformation of ψ at or near $0°$ is observed about 10% of the time for Pro (see section 2.4.3 for details) and extremely rare for other amino acids.

2.2.3 Constraint of dihedral angles

Ramachandran and his co-workers (1963, 1968) calculated the range of accessible ϕ and ψ angles, based on a rigid peptide bond with the dimensions shown in Fig. 2.6(a). The dimensional plot of ϕ and ψ, which is known as a Ramachandran plot, is an important way of representing the structural properties of repeating peptide conformations, single residues, or two successive residues.

Figure 2.8 represents the allowed main chain dihedral angles (ϕ, ψ) for Gly with no side chain or for the amino acids with a side chain, using the hard-shape peptide model with normal contact distances (—) and with the lower contact distances ($\cdots$) between atoms (Ramachandran and Sasisekharan, 1968). As shown in Fig. 2.8(a), Gly has a large and coherent allowed region comprising about 50% of the space. For

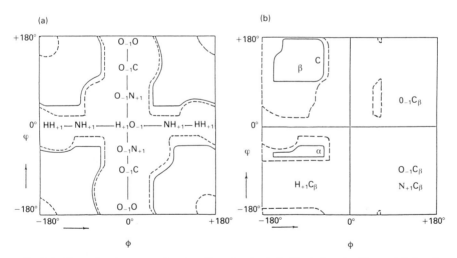

Fig. 2.8—Ramachandran maps showing allowed main chain dihedral angles. The peptide bond dimensions of Pauling *et al.* (1951a) are assumed. The graph is taken from Ramachandran and Sasisekharan (1968). (a) Map for Gly with no side chain, using the hard-sphere model. Prohibitive contacts are indicated in several areas. Atom names are given in Fig. 2.7. (b) Map for amino acids with a C_β atom, using the same hard-sphere model. Fully allowed conformations are outlined by a solid line and partially allowed conformations by a broken line. The conformations of right-handed α-helix (α), of β-pleated sheet (β) and of collagen (C) are indicated. The steric conflicts involving C_β atoms which restrict the large glycine area to the smaller area given here are indicated.

amino acids other than Gly the allowed conformational space is drastically reduced (Fig. 2.8b). Essentially similar plots of ϕ, ψ values are obtained for all amino acid residues (excluding Gly and Pro) of 12 proteins, as shown in Fig. 2.9. The ϕ, ψ values are estimated from the high-resolution X-ray crystal structures of the proteins. The major regions are the right-handed α-helical cluster in the lower left, near $-60°$, $-40°$; the broad region of extended β-sheets in the upper left quadrant centered around $-120°$, $140°$; and the sparsely populated left-handed α-helical region in the upper right near $+60°$, $+40°$, reflecting the high α-helix content of globular proteins. The asymmetry of the plot results from collisions with C_β. In each conformational region there exist significantly different structures: for example, parallel versus antiparallel β, widely varying degrees of β-sheet twist, and extended collagen-type helix lie within the β-region. The conformations of α-helix and β-sheets are described in sections 2.3.2 and 2.3.3.

Vacant regions in Fig. 2.9 are the conformations in which atoms are unfavorably placed close together within the peptide unit. The most inhibitive contact arises between H_{+1} and O_{-1} near $0°$, $0°$. The area across $\psi = 0°$ between the α- and β-regions, representing the contact between successive amide groups (Fig. 2.8a), should also be unfavorable based on a hard-sphere model. The area, however, is rather well

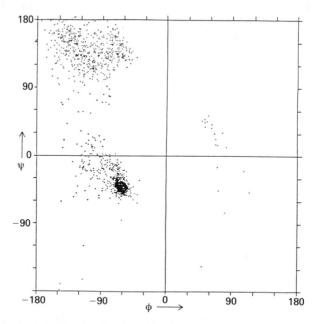

Fig. 2.9—Plot for ϕ, ψ values for all amino acid residues (excluding Gly and Pro) in the following proteins of highly refined X-ray resolution of 1.2–1.8 Å: dihydrofolate reductase (Bolin et al., 1982), erythrocruorin (Steigemann and Weber, 1979), flavodoxin (Smith et al., 1977), insulin (Dodson et al., 1979), myohemerythrin (Sheriff et al., 1987), ovomucoid, third domain (Papamokos et al., 1982), plastocyanin (Colman et al., 1978), pancreatic trypsin inhibitor (Wlodawer et al., 1984), RHE Bence-Jones protein (Furey et al., 1983), rubredoxin (Watenpaugh et al., 1980), Strep. griseus protease A (James et al., 1980), and scorpin (Cotton et al., 1979). (Redrawn from Richardson and Richardson, 1989.)

populated (Fig. 2.9). This may be partly due to stretching the bond angle τ and softening the amide nitrogen, which may, although rarely, permit the amide contacts.

Anderson and Hermans (1988) calculated the conformational energy for a dipeptide around a central C_β, which exhibits a rather good match of the observed ϕ, ψ distribution. This is rather surprising, because such a calculation neglects both the favorable and unfavorable effects of long-range interactions of the backbone as well as specific side-chain effects, which occur in proteins. This indicates that the optimum ϕ, ψ values and the permissible range for good long-range H-bonding as seen in α-helices and β-sheets are very close to the optimum and range for favorable dipeptide conformations. The ϕ, ψ plot for non-repetitive structure is presented in Fig. 2.10, which is comparable to Fig. 2.9, in spite of the absence of helix and sheet. This similarity is presumed to be due to the strong selection for the occurrence of right-handed α-helices rather than for any of the slightly different versions such as 3_{10}, π or left-handed α-helices. The structures of these α-helix versions are discribed in section 2.3.2. Such a fit of local and long-range preferences suggests that the influence of side-chain interactions is so variable that the effect is canceled out on the average. This would not necessarily be true for highly repetitive amino acid sequences, which produce a different range of conformations (Cook *et al.*, 1980; Rich and Crick, 1961). The structures of a cyclic repeat of two, three or four ϕ, ψ values

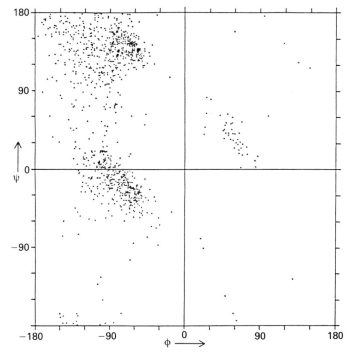

Fig. 2.10—ϕ, ψ plot for about 1000 residues in non-repetitive structure, excluding Gly and Pro, in highly refined X-ray crystal structures (about 2 Å resolution or better). (Redrawn from Richardson and Richardson, 1989.)

as reported by Cook et al. (1980) and Rich and Crick (1961) have never been found in the known globular protein structures.

In non-repetitive structures a clustering around the polyproline conformation (near $\phi = -60°$, $\psi = 140°$) is observed (Fig. 2.10). This ϕ, ψ plot is for the residues in non-repetititve structures excluding Gly and Pro. Therefore, like polyproline and 3_{10} conformations, about half of the latter involve Gly; some successive carbonyls in non-repetitive peptides may be approximately perpendicular to each other rather than parallel as in α-helix or antiparallel as in β-sheets. Most helices and β-sheets begin and end with a residue in one of the perpendicular conformations and a tight turn (see sections 2.3.2–2.3.4) requires two of them. The perpendicular conformations act as the punctuation between secondary structures or as the creases that fold a peptide chain into the elaborately folded conformation of a globular protein.

The Ramachandran maps are based on the assumption of a rigid peptide bond with the dimensions shown in Fig. 2.6(a). According to the model—a simple approach to steric hindrance in peptide structure—the contact between N and H_{+1} is completely inhibited (Fig. 2.8a). The potential energy map, which is an improved approach to steric hindrance (Brant et al., 1967; Nemethy and Scheraga, 1977; Ramachandran et al., 1966), is qualitatively the same as the hard-sphere map. However, a *bridge region* between the extended β-region (around $\phi = -120°$, $\psi = 120°$) and α-helix region (at $\phi = -60°$, $\psi = -60°$) is clearly demonstrated in the potential energy map and by the observed data on crystalline proteins (Fig. 2.9). This indicates an appreciable plasticity of peptide bond; i.e. the hindrance between N and H_{+1} is largely relaxed by a dipole–dipole interaction corresponding to a weak H-bond. It should be noted that slight hindrance can be circumvented by small deviations in bond and torsion angles as well as in bond lengths.

2.3 SECONDARY STRUCTURE

2.3.1 Factors determining protein structure

A polypeptide chain is folded into a three-dimensional structure by virtue of non-covalent forces and covalent bonds such as the disulfide bond. Protein folding

Table 2.3—Types of non-covalent forces important for protein structures (from Schulz and Schirmer, 1979).

Type	Example	Binding energy (kcal/mol)	Change of free energy water → ethanol (kcal/mol)
Dispersion forces	Aliphatic hydrogen	—C—H···H—C— −0.03	
Electrostatic interaction	Salt bridge	—COO⁻···H₃N⁺— −5	−1
	2 dipoles	$\overset{\delta+}{>}C=\overset{\delta-}{O}\cdots\overset{\delta-}{O}=\overset{\delta+}{C}<$ +0.3	
Hydrogen bond	Ice	>O—H···O< −4	
	Protein backbone	>N—H···O= −3	
Hydrophobic forces	Side chain of Phe		−2.4

strictly depends on the amino acid sequence of the polypeptide chain (Epstein *et al.*, 1963), although protein-like components (chaperonins) have been reported to play a role in protein folding (Horwich *et al.*, 1990).

Before describing the secondary and tertiary structures of proteins, the non-covalent forces determining protein structures are briefly presented in this section. The contributions of such forces to protein stability are described in sections 3.2.3–3.2.6. A survey of the forces is shown in Table 2.3.

2.3.1.1 *Dispersion forces*

The electrons moving around the nucleus in an atom generate an oscillating dipole. In a pair of atoms each dipole polarizes the opposing atom, which gives rise to an attractive force between the atoms. This attractive force is counterbalanced by the repulsion of the electronic shells.

The Linnard-Jones 6–12 potential for dispersion force and electron repulsion (Schulz and Schirmer, 1979) has a negative relative minimum at the distance R_m. Atoms are then weakly bound (E_m = attraction energy) at that distance. The energy contribution per atomic contact is very small (0.01–0.23 kcal mol^{-1}) (Momany *et al.*, 1974; Lifson and Warshel, 1968; Warshel *et al.*, 1970). Since the number of atomic contacts in a protein is large and contact energies add up linearly, the total attraction energy in a protein might be large.

2.3.1.2 *Electrostatic interactions*

Covalent bonds between different types of atoms cause an asymmetric bond eletron distribution. Thus, most atoms of a molecule carry partial charges. A neutral molecule, even of no net charge, contains dipoles or higher multipoles. These multipoles interact with each other according to Coulomb's law. Some typical examples for electrostatic interactions are given in Table 2.3. The interactions between completely ionized, charged atoms, which are usually called *salt bridges*, are also involved in the electrostatic interactions.

The actual range of electrostatic interaction is short. If the dipoles are aligned as found in α-helices and β-sheets, however, the weak long-distance interactions can add up. In α-helices, for instance, the dipoles form lines, canceling each other except for charges at both ends, which favors antiparallel helices rather than parallel ones.

2.3.1.3 *van der Waals potentials*

It is computationally convenient to combine the above three forces—the attraction and repulsion forces and electrostatic interactions—into a single, simplified potential function (Schulz and Schirmer, 1979). The resulting potential, called the van der Waals potential, depends only on the distance between the contact atoms. The definition of van der Waals contact distances between a pair of atoms is then derived from the van der Waals potential. In Talble 2.4 are given van der Waals radii as provided by Bondi (1964).

Lower limits for such distances between a pair of atoms have been given by Ramachandran and Sasisekharan (1968) and amount to about 75% of the equilibrium

Table 2.4 — Van der Waals radii (Bondi, 1964)

Type of atom	Radius (Å)
Aromatic H	1.0
Aliphatic H	1.2
O	1.5
N	1.6
C	1.7
S	1.8

distance R_m. These contact distances have been used to evaluate the steric hindrance of the hard-sphere model at the C_α-atom (Fig. 2.7).

The observed van der Waals contact distances for a pair of atoms are converted to the more general term, *van der Waals radii*, assuming that each distance is the sum of two atom radii. Since the resulting radii are averaged over all kinds of contact pairs, the content of van der Waals radii is only approximately valid.

2.3.1.4 Hydrogen bonds

The contact atomic distances between a pair of atoms can be expressed by the sum of van der Waals radii of the corresponding atoms, as described above. However, a number of atomic pairs involving a hydrogen atom as one of the partners exhibit a shorter contact distance than that calculated from the van der Waals radii. In such cases the hydrogen atom has a large positive partial charge and its contact partner has a large negative charge, which gives rise to an electrostatic interaction between dipoles (such as the peptide dipole, which puts a partial positive charge on the NH and a partial negative charge on the CO) or actual charges (as, for instance, in a Glu–Lys salt link). Sometimes the interactions involve the sharing of a proton.

The group on one side of the H-bond is the *donor* (D) (usually, in proteins, a nitrogen or a water but sometimes an OH), which has a hydrogen it can contribute to the bond. The other group is the *acceptor* (A), with accessible unpaired electrons (usually a CO or a water but sometimes an unprotonated N or the back side of an OH). The optimum distance for a strong H-bond is about 3 Å between D and A or 2 Å between H and A (Baker and Hubbard, 1984). The distance is a bit closer than the van der Waals constant and can be shorter than for a charged H-bond. The D–H–A angle is fairly critical, and the energy is optimal at 180° and almost zero near 120°. For the H–A–C angle (where C is the carbon to which A is attached), the optimum energy lies in the range 120–150°, but the interaction is still strong at either 180° or 90°. On the surface of a protein only a H-bond with very good geometry is useful, because of the competition with solvent (usually water) H-bonds. In the interior even a very long H-bond is better than none at all.

In the interior of proteins, indeed about 90% of all polar groups form H-bonds (Chothia, 1974).

2.3.1.5 *Hydrophobic interactions*

Thermodynamics of protein stability

Although a detailed description of the thermodynamics for protein stability is presented in Chapter 3, the thermodynamic treatment of protein unfolding is briefly presented here to explain hydrophobic interactions in proteins.

Protein stability depends on the free energy change for protein unfolding (between the folded and unfolded states of proteins). The free energy change is thermodynamically expressed by the equation

$$-RT \ln K = \Delta G = \Delta H - T\Delta S \tag{2.1}$$

where R represents the Avogadro number; K, the equilibrium constant (number of polypeptide chains in folded state (N)/number in unfolded state (U); ΔG, the free energy change ($G_N - G_U$); ΔH, the enthalpy change ($H_N - H_U$); and ΔS, the entropy change ($S_N - S_U$).

In most biological reactions the changes of volume and volume energies can be neglected, so that Gibbs free energy equals Helmholtz free energy and binding energy equals binding enthalpy. ΔH in the above equation corresponds to the binding energy; i.e. dispersion forces, electrostatic interactions, van der Waals potentials and H-bonds, as described above. The hydrophobic interactions as given in Table 2.3 are related to the entropy term, ΔS.

It is apparent that proteins are more stabilized as the negative value of ΔG becomes larger; i.e. the larger the negative value of ΔH and the smaller the negative value of ΔS. In other words, proteins are stabilized when the binding energy is increased or binding interactions are strengthened and when the difference in entropy between the folded and unfolded states becomes smaller.

Entropic effect

Native proteins necessitate water to maintain their accurately folded states. Thus, considering the interactions between protein and water molecules equation (2.1) should be as follows:

$$\Delta G = \Delta H_{chain} + \Delta H_{water} - T(\Delta S_{chain} + \Delta S_{water}) \tag{2.2}$$

Here, the subscript 'chain' represents the thermodynamic term relating to polypeptide chain and 'water' the term relating to water. The interaction energy between polypeptide chain and water is involved in the ΔH_{chain}.

When mineral oil consisting of hydrocarbons is added to water, it easily forms into an oil drop on the water surface, rather than a monodisperse solution of oil in water. The former is called the *N-state*, and the latter the *U-state*. In equation (2.2) (ΔH_{oil} and ΔS_{oil} instead of ΔH_{chain} and ΔS_{chain}), ΔS_{oil} is negative, while ΔH_{oil} is positive, because of stronger van der Waals forces in the N-state than the U-state. Electrostatic interactions between the dipoles of water and the dipoles in hydrocarbons induced by the polar water should be additionally considered (N-state), besides the dispersion forces between oil molecules (U-state) (Kauzmann, 1959). ΔS_{water} is positive, favoring the N-state. This is explained by a greater ordering of water in the U-state than the N-state, due to a larger oil–water surface area in the U-state. Water molecules assume a locally ordered arrangement around oil molecules (a cage-like structure) (Kauzmann, 1959). Even with this quasi-solid structure, water forfeits some hydrogen bonds, giving rise to a negative ΔH_{water} which favors the N-state.

In summarizing the above, ΔH_{oil} and ΔS_{oil} favor the U-state, whereas ΔH_{water} and ΔS_{water} favor the N-state. Since a phase separation between oil and water is indeed observed, the latter contributions surpass the former ones. Moreover, ΔH_{water} is small. Consequently ΔS_{water} drives the oil molecules to aggregate into an oil drop. The ΔS_{water} is called *entropic effect* or *hydrophobic forces*.

Entropic effect on polypeptide chains

Similarly to the oil–water system, an extended, unfolded polypeptide chain (U) is folded into a three-dimensional structure (N) through an entropic effect derived from the interactions between hydrophobic, non-polar residues and water (Kauzmann, 1959).

The contribution of entropic effect to protein stability can be qualitatively explained as follows (Schulz and Schirmer, 1979). As described in the previous section, for non-polar residues the magnitudes of both ΔH_{chain} (positive) and ΔH_{water} (negative) are small, while ΔS_{water} is positive and large. For all residues of the polypeptide chain ΔS_{chain} is negative.

Polar residues form H-bonds with water when exposed to it and hydrogen bonds with each other in the protein interior. This means that one H-bond with water is lost and half a hydrogen bond is formed per polar residue. Therefore, ΔH_{chain} is positive, while ΔS_{water} is negative. The magnitude of ΔH_{chain} is smaller than that of ΔH_{water}, because of more van der Waals contact in the tightly packed protein interior (see below) than at the chain–water contact. This contribution favors the N-state. As water molecules are slightly ordered around polar residues, ΔS_{water} is positive, but small. Taken together, ΔS for polar residues is around zero.

Thus, it can be concluded that ΔS_{water} from non-polar residues gives the largest single contribution to protein stability. The large ΔH_{water} contribution for polar residues is essentially canceled by ΔH_{chain}.

Molecular packing

The hydrophobic side chains aggregate to form the hydrophobic interior of the protein. Some non-polar side chains, however, are expelled from the molecule to form a rigid or compact protein conformation. The exterior non-polar residues occupy about half the surface area, destabilizing the protein (Mozhaev *et al.*, 1988). The compactness of a protein or high packing density in the protein interior is defined as the ratio of the bare molecular volume (the van der Waals envelope of the molecule or the van der Waals radii of the surface atoms) to the volume actually occupied in space. Since local packing density may vary over the whole protein, the estimation of local packing is more promising for revealing features of structural significance. The sum of all local packing densities is the average packing density of the protein. Local packing densities were first determined by Richards (1974, 1977). The observed local packing densities of proteins are 0.68–0.82, and low packing densities are found in active centers, which may reflect the flexibility of active centers of enzymes (Dixon and Webb, 1979). The average packing density of a protein is around 0.75. Since crystals of small molecules which are held together by van der Waals forces have packed densities of 0.70–0.78, proteins can be considered to be as densely packed as small molecules in van der Waals crystals.

The protein manages to bring almost all interior polar groups into H-bonds, although the bonds pose geometric constraints which have to be reconciled with the geometric constraints of dense packing. Such conformational optimization is reflected in the observation that evolutionary changes in the protein interior are much rarer than changes at the surface. In addition, internal changes tend to compensate each other, e.g. the change Ile to Val may be accompanied by the adjacent Gly to Ala (Wyckoff, 1968).

2.3.2 Helices

There are two patterns of main chain hydrogen bonding with repeating values of ϕ, ψ angles. The α-helix, the quintessential aspect of protein structure, involves repeating patterns of local H-bonding. The β-structure (or β-sheet) involves repeating patterns of H-bonds between distant parts of the backbone polypeptide chain.

α-Helix

The α-helix was first postulated by Pauling *et al.* (1951) and confirmed by the X-ray analysis of hemoglobin (Perutz, 1951).

The α-helix is formed by repeated H-bonds between the CO of residues n and the NH of residues $n+4$, with repeated ϕ, ψ values near $-60°$, $-40°$ (Fig. 2.11). The helices found in natural proteins are right-handed. This is due to the cumulative effect of a moderate energy difference for each residue and mostly to the collision of C_β with the following turn of a left-handed helix.

The repeat, or pitch, of an ideal α-helix is 3.6 residues per turn, and for each pitch the rise per residue along the helix axis is 1.5 Å or 5.4 Å per turn. Real helices which are abundantly present in globular proteins match this value. A difference between 3.5 and 3.7 residues, however, is within the common range of variation. Such difference produces a small change in ϕ, ψ angles, but considerably affects side-chain packing (Banner *et al.*, 1987). The average length of α-helix is about 17 Å, which corresponds to 11 residues or three helix turns (Srinivasan, 1976). α-Helices provide rather stable rods in protein structures. The conformational parameters for α-helices are shown in Table 2.5.

The H-bonds in an α-helix are nearly parallel to the helix axis, and the COs all point toward the C-terminal end (Fig. 2.11). Since each peptide is slightly slanted, all the oxygens point outward to some extent (Fig. 2.12a). Fig. 2.12(b) shows the α-helix viewed from the N-terminus, which indicates almost square corners at the α-carbons. Appreciable offset, however, is observed at each successive turn. The β-carbons make a clockwise pin wheel shape with C_β nearly in the plane of the preceding peptide. The C_α—C_β bonds tilt somewhat in the N-terminal direction and are rather close to the backbone of the preceding turn.

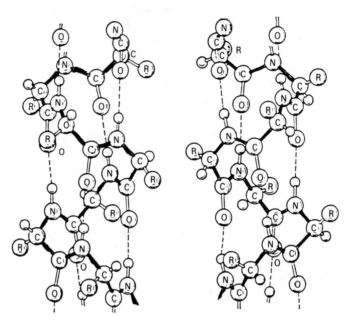

Fig. 2.11—Conformation of α-helix. (a) Left-handed α-helix (α_L). (b) Right-handed α-helix (α_D). α_D is more stable than α_L. Broad lines indicate α-carbon chains. Broken lines represent H-bonds.

Table 2.5—Structural parameters of helices and sheets occurring in proteins

Linear group	Observed	Residues per turn n and chirality[a]	Rise per residue d(Å)	Radius of helix r (Å)
Planar parallel sheet	Rare	±2.0	3.2	1.1
Planar antiparallel sheet	Rare	±2.0	3.4	0.9
Twisted parallel or antiparallel sheet	Abundant	−2.3	3.3	1.0
3_{10} Helix	Small pieces	+3.0	2.0	1.9
α-Helix (right-handed)	Abundant	+3.6	1.5	2.3
α_L-Helix (left-handed)	Hypothetical	−3.6	1.5	2.3
π-Helix	Hypothetical	+4.3	1.1	2.8
Collagen-helix	In fibers	−3.3	2.9	1.6

[a] Plus and minus correspond to right- and left-handed helices, respectively (from Schulz and Schirmer, 1979)

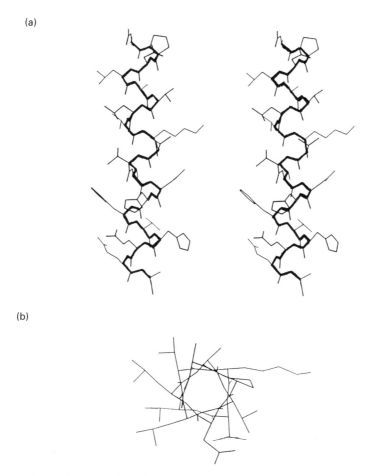

Fig. 2.12—Stereo view of an α-helix from the side (a) and from the end (b) in myohemerythrin (Sheriff *et al.*, 1987). Broad lines represent α-carbon chains. (Redrawn from Richardson and Richardson, 1989.)

3_{10} helix

The other type of helix which often occurs is the 3_{10} helix. The 3_{10} helix has an H-bond between the CO of residue n and the NH at residue $n+3$, and ϕ, ψ values near $-70°, -5°$ (Table 2.5). The 3_{10} helix is more tightly wound than the α-helix. As the 3_{10} helix has the integral number of residues per turn (3.0 as opposed to 3.6 for the α-helix), the α-carbons on successive turns of the helix line exactly with each other. This induces the tilted H-bonds relative to the helix axis. For the α-helix with the non-integral pitch of 3.6 residues per turn, the α-carbons do not line up, but as described above a CO on one turn and an NH on the next turn line up and the H-bonds are parallel. As compared to the α-helix, the side-chain packing of the 3_{10} helix is rather unfavorable. The position of the α-helix in the center of an allowed region of the ϕ, ψ map (Figs 2.8a and 2.9), with the 3_{10} helix at the edge of the map, indicates less stability of the 3_{10} helix relative to the α-helix. The well-packed side chains for the α-helix, together with the aligned dipoles of H-bonds and van der Waals attractions across the helix axis allowed by the helix radius, may stabilize its structure more than the 3_{10} helix. The energetically disadvantageous geometry of the 3_{10} helix accounts for its rare occurrence in proteins (Hendrickson and Love, 1971; Hendrickson *et al.*, 1973). The major importance of the 3_{10} helix is that it frequently forms the last turn at the C-terminus of an α-helix.

Collagen is a superhelix formed by three parallel, extended left-handed helices (Traub and Piez, 1971). The peculiar conformation of collagen, however, is beyond the scope of this book. The outline of its conformation is referred to in Schulz and Schirmer (1979) and Creighton (1973), or in textbooks such as Lehninger (1981).

2.3.3 β-Sheets

The other common type of repeating secondary structure is the extended β-sheet (Fig. 2.13). Parallel (Fig. 2.13b) or antiparallel (Fig. 2.13a) polypeptide chains are H-bonded with ϕ, ψ values of $-120°, 140°$. Figure 2.14 shows a schematic presentation for the two types of β-sheet. The almost fully extended β-corformation is characterized by the following perpendicular directions: (1) the N-terminus to C-terminus direction of the backbone; (2) the plane of the peptides, in which close pairs of H-bonds are formed alternately to the left and to the right; and (3) the direction in which the side chains extend, alternating one up and one down.

As shown in Fig. 2.13, in an antiparallel sheet the H-bonds are perpendicular to the strands and alternate between a closely spaced pair and a widely spaced pair. The H-bonds in a parallel sheet are evenly spaced but alternately slant forward and backward.

The side chains which are approximately perpendicular to the H-bond plane alternate from one side to the other. For an antiparallel β-sheet one side chain is buried in the interior and the other side is exposed to solvent. Thus, the amino acids tend to alternate hydrophobic and hydrophilic. On the other hand, parallel sheets are usually buried in both side chains, so that the central amino acid sequences are highly hydrophobic, and hydrophilicity concentrates at the ends.

The average length of sheets in globular proteins, which contain about 15% sheet structure (Chou and Fasman, 1974a), is about six residues. The distribution ranges from two to 11 residues (Sternberg and Thornton, 1977). Most sheets contain less than six strands, and the width of the six-stranded sheet is about 25 Å, which corresponds to typical domain diameters. Domain structures are described later. Although no preference of parallel and antiparallel sheets is observed, parallel sheets with less than four strands are rare.

Twist structure

The most interesting and characteristic feature of β-structure is the twist observed both for the strands and for whole sheets (Chothia, 1973). The planar, antiparallel sheet postulated by Pauling and Corey (1951) is rare and is observed in glutathione reductase, for example (Schulz *et al.*, 1978). The planar sheet is illustrated in Fig. 2.14(a). The planes of the peptides (and also of H-bonding) (Fig. 2.13) twist in a right-handed sense along the strand direction; i.e. a left-handed twist when viewed along the sheet

2.3 Secondary Structure

(a)

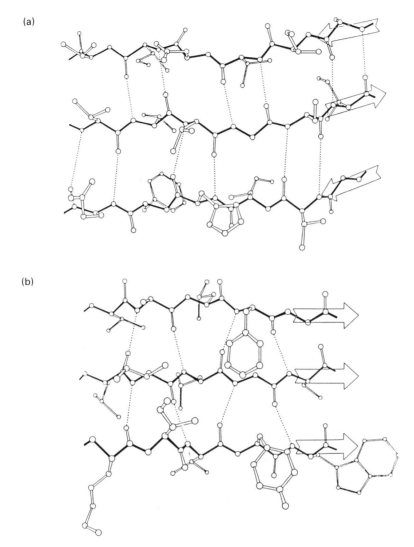

(b)

Fig. 2.13—(a) Antiparallel β-sheet in Cu,Zu superoxide dismutase (Tainer *et al.*, 1982). The figure shows the alternately narrow and wide pairing H-bonds and the side chain alternation above and below the plane of the sheet. (b) Parallel β-sheet in flavodoxin (Smith *et al.*, 1977), indicating the evenly spaced but alternately slanting H-bonds. Both are redrawn from Richardson and Richardson (1989).

plane perpendicular to the strands. Twist structures are illustrated in Fig. 2.15, and Fig. 2.16 shows one of the most twisted sheets (30° per residue). Taking twist conformation the close contact of side chains opposite one another on neighboring strands are partly relieved. If the strands go left and right (Fig. 2.13), upper-right-to-lower-left diagonal pairs facing the observer can contact each other, while pairs on the other diagonal are widely separated.

The β-sheets of all the types mentioned above can be distributed in a wide region—the right of the diagonal $n = 2$ line, on the Ramachandran plot (Fig. 2.15a;

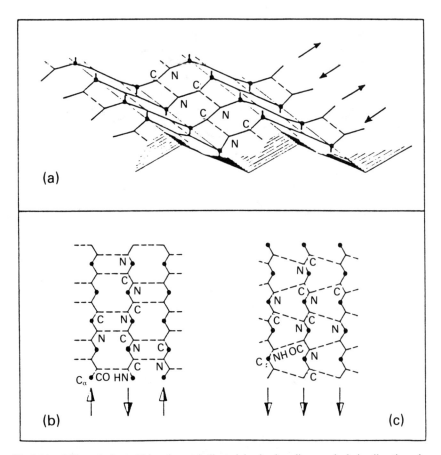

Fig. 2.14—β-Pleated sheet. H-bonds are indicated by broken lines and chain directions by arrows. C_α-atoms are marked by dots. (a) The pleated structure of a non-twisted, planar antiparallel β-sheet. The directions of the C_α—C_β bonds are indicated by short bars on the C_α atoms. They are perpendicular to the sheet. (b) H-bond pattern of an antiparallel three-stranded β-sheet. (c) H-bond pattern of a parallel three-stranded β-sheet. (Redrawn from Schulz and Schirmer, 1979.)

see also Fig. 2.8). The average conformation, an overall twist of the sheet, will be at the center of this region.

In addition to the commonest right-handed twist, β-sheets can be *curled* along the strand direction or *arched* perpendicular to it. Arching, which can be seen, for example, in flavodoxin (Smith *et al.*, 1977), does not necessitate any changes in strand conformation. Moderate arching can be accommodated by the H-bonds. Curling, on the other hand, involves no change in H-bond geometry but an alternating perturbation of ϕ, ψ values (Chothia, 1983).

Curl is visualized as the β-ribbon, and is most common for two-stranded sheets as in Fig. 2.16. In Fig. 2.17 the ribbon conformation is shown schematically. The amino acid residues of one side of the peptide chain are between narrow, concave pairs of H-bonds, and those of the other are between wide and convex pairs. If the

2.3 Secondary Structure

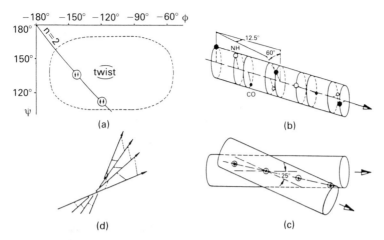

Fig. 2.15—Twist of β-pleated sheets. (a) Part of φ, ψ map with dihedral backbone angles for non-twisted parallel (↑↑), non-twisted antiparallel (↑↓), and left-handed twisted β-pleated sheet. The dashed line is the potential curve of 1 kcal/mol. (b) A single strand of a twisted sheet is twisted by itself. The backbone course is indicated by a dashed line. The directions of carboxyl group (●) and amide group (○) are indicated. (c) Two parallel chains, which are twisted as sketched in (b), can be superimposed so as to form H-bond (⊙). For this purpose they have to be tilted by 25°. (d) Schematic drawing of a parallel sheet with the usually observed left-handed twist. (Redrawn from Schulz and Schirmer, 1979.)

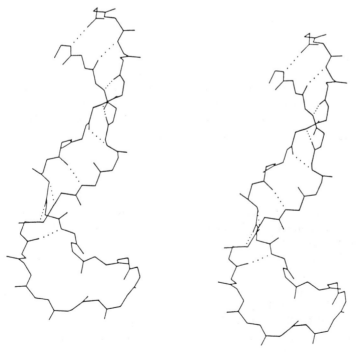

Fig. 2.16—A strongly twisted, two-stranded antiparallel β-ribbon in lactate dehydrogenase (White et al., 1976). (Redrawn from Richardson and Richardson, 1989.)

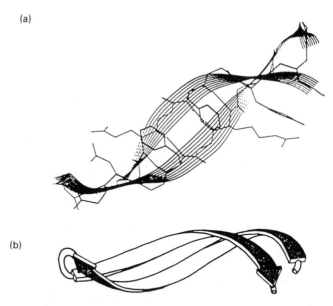

Fig. 2.17—The *sideness* of a two-stranded β-ribbon. (a) The detailed conformation observed in pancreatic trypsin inhibitor (Wlodawer *et al.*, 1984) is shown. The side chains on one side are between a narrow pair of H-bonds and on the other side between a wide pair. (b) A schematic drawing of the same twisted pair. The hydrophobic, narrow-pair, concave side is shown white, and the exposed, wide-pair, convex side is shown black. The view is from the concave side and the arrow directions are clockwise. (Redrawn from Richardson and Richardson, 1989.)

amino acid residues on the H-bonded chains—narrow and concave—are hydrophobic, especially large ones which interact strongly with each other, such β-ribbon will promote a compact, stable structure.

The parameters such as twist, pleat, arch and curl, sheet vary significantly for β-sheet, but they are accommodated within a regular network of β-sheet H-bonding. Local disruption of H-bonding in β-structure which is called *β-bulge* is frequently observed (Richardson *et al.*, 1978). Bulges are common in antiparallel β-structure, and rare in parallel β-sheet.

2.3.4 Non-repetitive structures

In addition to the repetitive structures such as α-helices and β-sheets, non-repetitive but well-ordered, compact and stable structures are observed in globular proteins. They are turns, connections and compact loops.

Venkatachalam (1968) found a tight turn consisting of four successive residues with H-bond between the CO of the first residue and the NH of the fourth residue. The most common type of turn is the *common* turn, formerly called type I turn or reverse turn I. The second residue of the turn has the ϕ, ψ values of α-helix and the third one those of 3_{10} (Fig. 2.18a). Any residues are permitted for the turn. The second most common type is the *glycine* turn, formerly type II (Fig. 2.18b). The second and third residues are, respectively, in poly-Pro and in left-handed 3_{10} conformations. An $L3_{10}$ conformation is favorable only for Gly, so that about 60% of the glycine turn

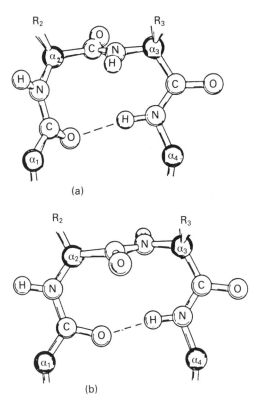

Fig. 2.18—Schematic drawings of common turn (a) and glycine turn (b), in which R_3 is most often G side chain (Gly).

has Gly in position 3. Glycine turns have all four α-carbons and the H-bond almost in a plane. The common turn, on the other hand, is non-planar, its four α-carbons make a dihedral angle of about +45° and the CO of residue 1 is almost perpendicular to the turn plane. One classic role of turns is the connect two consecutive antiparallel β-strands in a hairpin connection. The right-hand twist of β-strands would form a negative dihedral angle at the end of the hairpin, typically about −45°. The opposite twists, when connected by the common turn, would fight each other (Sibanda and Thornton, 1985). As shown in Fig. 2.19, the inverse common turn, which is the mirror image of the common turn, has a dihedral angle of −45° for the four carbon atoms making the turn. This is one of the ways of solving the incompatible twist. Bulged hairpins are also commonly used for that purpose (Richardson, 1981).

There exist many varieties of loops and connections with non-repetitive structures. Loops have close termini, good internal packing interactions and many side-chain–main-chain H-bonds (Leszcaynski and Rose, 1986). In contrast to secondary structures, which are defined and stabilized by main-chain H-bonds, loops are defined and stabilized by side chains. The side chains make loops compact, and they provide H-bonds to stabilize the loop conformations. A given side-chain type has a strong preference for H-bonding at a certain distance and position along the backbone, which influences loop configurations (Baker and Hubbard, 1984). Compact loops can be

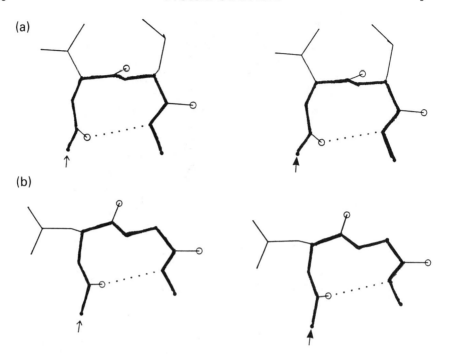

Fig. 2.19—Stereo drawings of (a) common turn in thermolysin and (b) inverse common turn in prealbumin (Blake *et al.*, 1978). (Redrawn from Richardson and Richardson, 1989.)

among the most highly ordered parts of a protein, although they are usually on the protein surface and sometimes disordered.

Connections between α and α (Efinov, 1982), between β and α and between α and β (Edwards *et al.*, 1987) have been reported.

Loops in globular proteins are proposed to constitute an additional secondary structure (Leszczynski and Rose, 1986). These so-called ω-loops were previously classified as random coil. The segment length of ω-loop is between six and 16 residues.

2.3.5 Supersecondary structures

Relating to the connections between the ordered structures, as described above, the secondary structures are often aggregated to form so-called supersecondary structures.

Coiled-coil α-helix

A clear example of supersecondary structure is the coiled-coil α-helix, which is found in fibrous proteins such as α-keratin (Parry *et al.*, 1977) and tropomyosin (Caspar *et al.*, 1969). This structure, although in short pieces, is also observed in globular proteins such as bacteriorhodopsin (Henderson and Unwin, 1975) and tyrosyl transfer RNA synthetase (Irwin *et al.*, 1976). The globular proteins contain α-helices packed in an approximately parallel or antiparallel way.

As shown in Fig. 2.20, the coiled-coil α-helix is formed by meshing two adjacent α-helices in a left-handed coil. The contact line arises along which side chains from both helices are regularly meshed. The ϕ, ψ values of coiled-coil α-helix slightly deviate

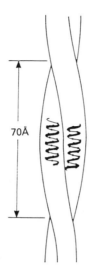

Fig. 2.20—Left-handed supercoiled helix with a repeat of 140 Å. Both helices are parallel. (Redrawn from Schulz and Schirmer, 1979.)

from those of the standard α-helix. Since meshing of side chains provides for a broad and intimate contact between helices, the coiled-coil α-helix is energetically favorable. If the meshed side chains are hydrophobic the free energy of this structure is especially low.

βξβ-Unit

Another supersecondary structure is the $\beta\xi\beta$-unit, which is two parallel strands of a β-sheet with a connection ξ in between. If ξ is a non-regular chain, the unit is called $\beta c\beta$; if α-helix, $\beta\alpha\beta$; and if β sheet, $\beta\beta\beta$. The $\beta\xi\beta$-units observed in globular proteins have right-handed twist conformations.

A combination of two consecutive $\beta\alpha\beta$-units is frequently found as the Rossmann fold (Rao and Rossmann, 1973).

2.4 TERTIARY STRUCTURE

Tertiary structures are largely determined by the packing of α-helices and/or β-strands (Levitt and Chothia, 1976; Chothia *et al.*, 1977; Chothia and Janin, 1982; Richardson, 1976, 1977; Sternberg and Thornton, 1976, 1977; Richmond and Richards, 1978). Such associations of secondary structures satisfy the H-bonding requirements of buried main-chain nitrogen and oxygen atoms, while shielding a substantial fraction of the non-polar atoms from solvent (Chothia and Janin, 1975). The supersecondary structures are the main components of the domain which, when assembled, constitutes the three-dimensional structures of proteins.

2.4.1 Structural domains

Electron density maps calculated in the course of *X*-ray analysis of globular proteins showed that most proteins, even of smaller molecular sizes, consist of several globular parts which are connected loosely with each other. The parts denoted by *structural domains* have been defined in a number of different ways (Wetlaufer, 1973; Rose, 1978; Richardson, 1981).

Structural domains are supposed to be the folding units of a polypeptide chain. This image arises from the structural analysis of chymotrypsin, because the protein contains 13 water molecules between the two domains which make up the protein (Birktoft and Blow, 1972). Presumably the two domains fold separately; the folding is not necessarily complete. The water molecules are trapped during the subsequent association of the domains.

Most of the larger globular proteins can be split into several structural domains of 100–150 amino acid residues, which corresponds to a globule of about 25 Å diameter. One large globule has a much smaller ratio of surface to volume than the sum of those of the individual small ones. Thus the large globule can form a large hydrophobic core and many internal hydrogen bonds, both of which are energetically favorable. This indicates a preference for a large globule. A small folding unit is presumably necessary to keep the folding process simple.

According to Wetlayfer (1973) a polypeptide chain can be divided into consecutive pieces which belong to consecutive domains, and the residues that are far apart along the chain are at a great geometric distance. Thus, the domain structure exhibits a high degree of neighborhood correlation. Neighborhood correlation can be understood kinetically and thermodynamically (Schulz and Schirmer, 1979). Sternberg and Thornton (1977) showed that consecutive antiparallel β-sheets have a strong tendency to be adjacent

2.4.2 Structural types

Tertiary structures of globular proteins can be described at the level of a single domain, classified into the following major groups according to the types of secondary structures involved in proteins (Richardson, 1981).

All α-type structure

Naturally occurring proteins or domains consisting only of α-helices have either an antiparallel or perpendicular arrangement of helices. This is probably due to a preference for near-neighbor dipoles (Hol *et al.*, 1978) or for near-neighbor helix interactions during folding (Richardson, 1981).

The simplest type of antiparallel helix structures is a cylindrical bundle or cluster of four helices with up-and-down, near-neighbor connectivity (Argos *et al*;, 1977; Weber and Salemme, 1980). Diagrams of some arrangements for four-helix clusters are given in Fig. 2.21. In addition, there are a few three-helix bundles, and also antiparallel clusters of five or more helices with at least one non-near-neighbor connection. Figure 2.22 presents the ribbon drawing for the above-described all-α structures. The structures with one or more connections between non-near-neighbor α-helices (or β-strands) are called Greek bundles or barrels (Richardson and Richardson, 1989).

Another type of all-α protein is the successive helix pairs which are connected perpendicularly at their sequence-adjacent ends. The great majority of adjacent helix

2.4 Tertiary Structure

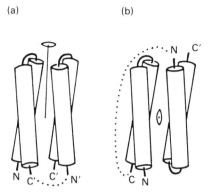

Fig. 2.21—Diagram of two alternative arrangements for helix-pair dimers. Dotted lines show the two types of connection of two helix-pair dimers. (a) The dimers are related by a lengthwise twofold axis, and they are joined by a short connection into an up-and-down helix bundle. (b) The dimers are related by a crosswise twofold axis, and they are joined by a long cross-over connection.

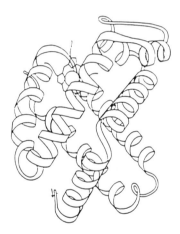

Fig. 2.22—Ribbon drawing of the β-subunit of hemoglobin (Fermi et al., 1984). This is an example of a Greek key helix bundle with one non-near-neighbor connection. (Redrawn from Richardson and Richardson, 1989.)

pairs are arranged with the two helix axes intersecting near their common ends, as though they had formed by bending a single, longer helix (Richardson and Richardson, 1988). Figure 2.23 illustrates one such type of conformation.

Parallel α/β structure

This structure consists of α-helices and β-strands. The two ordered structures alternate along their sequences and are organized in multiple layers, as shown in Fig. 2.24(a). Parallel β-sheets are inside and layers of helices cover them. The structures are constructed from the supersecondary structure, right-handed $\beta\alpha\beta$ (section 2.3.5). Thus,

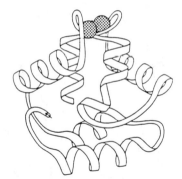

Fig. 2.23—An example of proteins containing offset perpendicular helix corners observed in carp Ca-binding protein (Kretsinger and Nockolds, 1973). Shaded balls are Ca ions. (Redrawn from Richardson and Richardson, 1989.)

(a)　　　　　　　　　　　　　　　(b)

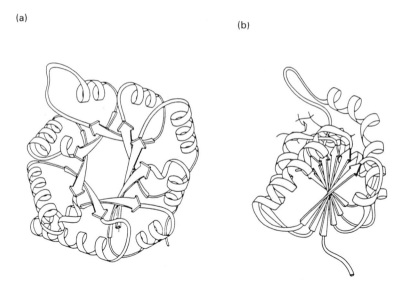

Fig. 2.24—Two types of parallel α/β. (a) Ribbon drawing of the classic example of a single-wound parallel β-barrel in triose phosphate isomerase (Banner *et al.*, 1975). (b) A doubly wound parallel β-sheet, nucleotide-binding domain, in lactate dehydrogenase (White *et al.*, 1976). (Redrawn from Richardson and Richardson, 1989.)

parallel α/β structures are overwhelmingly right-handed. More than a dozen of these structures are reported with eight parallel strands in the central β-region and a concentric cylinder of eight α-helices formed by right-handed cross-over connections (Fig. 2.24a).

A slightly different organization of parallel α-helices and β-sheets is the doubly wound parallel β-sheet with cross-over right-handed connections (Fig. 2.24b). The conformation in Fig. 2.24a is a single wound parallel β. The structures as shown in Fig. 2.24(b) (Rossmann fold) are commonly found for globular proteins, especially for dehydrogenases.

2.4 Tertiary Structure

Antiparallel β-structure

The third major group is antiparallel β-sheet proteins, forming barrels or single sheets. The most common structures of this group are antiparallel barrels with one or more connections that cross the top or bottom of the barrel, often skipping two intervening strands. The simplified layout of the structure is shown in Fig. 2.25.

Barrels with more than six strands have a flattened or elliptical cross-section unless there is something bound in the central hydrophobic core (Fig. 2.26). Taking such structure the side chains can pack well between opposite β-sheets. Because the strands twist, the direction of flattening twists, often by close to 90° from top to bottom. These structures are described as β-sandwiches (Chothia and Janin, 1981).

The other common organization for antiparallel β-structure—open-faced sandwich—is a single sheet, twisted but not closed into a cylinder, with helices and/or loops packed against one side. In small proteins the other side of the sheet is exposed

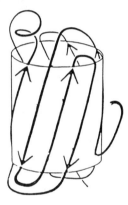

Fig. 2.25—Drawing of a Greek key barrel appearing in a subunit of prealbumin (Blake *et al.*, 1978). (Redrawn from Richardson and Richardson, 1989.)

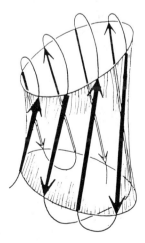

Fig. 2.26—Illustration of the way in which the direction of barrel flattening (the elliptical cross-section) twists from top to bottom. (Redrawn from Richardson and Richardson, 1989.)

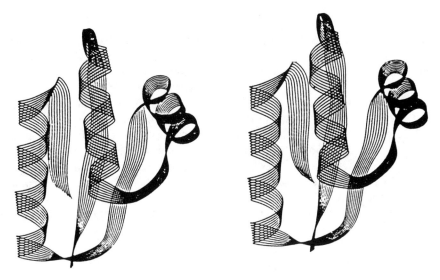

Fig. 2.27—Stereo view of β-ribbon in ribosomal protein L7/L12 fragment (Leijonmarck and Liljas, 1987). A small open-face β-sheet with an N-centered overhand topology is seen. (Redrawn from Richardson and Richardson, 1989.)

to solvent, while in larger proteins it may contact other domains or subunits. The structures of the open-faced sandwich type provide the covering helices or loops within the β-sheet region or immediately adjacent to it, perhaps because of the hydrophobicity of isolated β-sheet peptides. Three types of arrangement for three-stranded sheet are given in Fig. 2.27. In some cases the three-stranded sheets are apparently combined to form larger sheets.

2.4.3 Roles of individual amino acids

A polypeptide chain is non-catalytically folded into the characteristic three-dimensional structure. This self-assembly capacity of a polypeptide chain is exclusively derived from the properties of the amino acid residues lined up along the chain. The interactions of the residues with the backbone, with solvent and with one another determine the thermodynamically stable structure. The properties relating to the interactions are the size, shape, hydrophobicity, charge, H-bonding and degree of freedom of the side chains. As described previously, and as will be discussed in Chapters 3 and 7, the properties of side chains play important roles in stabilizing proteins as well as in forming a three-dimensional structure of proteins. In the following, then, the roles of the individual amino acids are briefly considered.

Glycine

Gly has no side chain but only a hydrogen. The lack of C_β means that Gly has a symmetrical conformation of rather free rotation. Thus Gly can be located at positions in repetitive and non-repetitive structures which disfavor other amino acids.

One of the most common endings of an α-helix necessitates the left-handed 3_{10} or L_α conformation (Schellman, 1980). This unusual conformation might endow

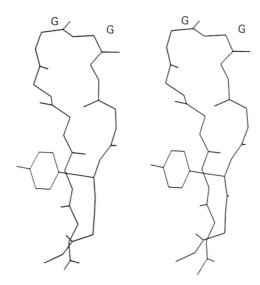

Fig. 2.28—Stereo view of a tight β-hairpin with glycines at positions 2 and 3 of an inverse common turn. The structure is from actinidin (Baker and Dodson, 1980). (Redrawn from Richardson and Richardson, 1989.)

Gly with an α-helix breaker. In addition, this kind of helix ending with Gly sends the backbone off in a new direction (Fig. 2.28).

As noted in section 2.3.4, the turn structures such as Gly, inverse Gly and inverse common turns contain Gly. A tight hairpin between consecutive β-strands requires an inverse turn, because of the twist of strands (Sibanda and Thornton, 1985). Therefore, most tight hairpins possess at least one Gly.

The rather free rotation of Gly and its broader range of accessible conformations result is another role for protein conformation, especially stability. From entropic consideration, Gly actually favors the unfolded state. For example, protein stabilization by replacing Gly with Ala, if there is no conformational or steric restriction against the change, is partly brought about by lowering the entropy of the unfolded state (Hecht et al., 1986).

Alanine

Ala has one methyl group as a side chain, which is the smallest, non-polar residue. Thus Ala is located either in the interior or at the surface of proteins. The amino acid can form the stablest H-bonded structure, and is the third strongest α-helix former (section 2.7), according to Chou and Fasman (1974b). Als is prevalent in α-helix, and exhibits a slight preference for either end relative to the middle or for outside versus inside (Chou and Fasman, 1974a).

A β-sheet prefers to be covered by larger side chains for stability (section 2.3.3). Turns prefer more hydrophilic residues, and non-repetitive structure favors amino acids with side chains of H-bonding capability (section 2.3.4). An α-helix, on the other hand, is completely satisfied by the helical H-bonds. In contrast to any other helix-forming residues, Ala exhibits only a slight preference for either end relative to the middle or for outside versus inside (Richardson and Richardson, 1988).

Sixty per cent of inside antiparallel β-barrels are occupied by the large, hydrophobic side chains such as Val, Ile, Leu and Phe (Chothia and Janin, 1981). These chains, however, cannot be closely packed by the scattered short side chain Ala.

Ala is important in α-helix formation early in protein folding (shoemaker et al., 1987).

Valine, isoleucine, leucine and methionine

The aliphatic, non-polar side chains branched at C_β are mostly (85–90%) located in the protein interior (Rose et al., 1985). These variously shaped side chains are well fitted together and compactly packed in the interior, although the side chains allow constant, rapid juggling of the hydrophobic core. The relation of aliphatic amino acids to protein stability was examined by Ikai (1980). Met provides a very flexible hydrophobic side chain, in contrast to other aliphatic non-polar chains.

On β-sheets the neighboring side chains, both alternate ones along a strand and those on adjacent strands, contact each other and form a continuous layer. Side chains from other than those along the sheets pack against the outside of the layer or penetrate it partially, but not to the backbone. The hydrophilic chains on the solvent-exposed face of an antiparallel β-sheet are less bulky and the sheet is often convex, so that solvents penetrate as far as the backbone in between them, and break the continuous layer. On buried faces the backbone is fully covered by its own side chains.

The contacts between neighboring side chains on α-helices are rare, so that solvent can penetrate as far as the backbone on solvent-exposed faces. In helix–helix contacts the side chains interdigitate exclusively.

Branching at C_β of Val and Ile (and also Thr) puts stronger constraints on the χ_1 of the amino acids. The values of χ_1 for all the side chains are around $-60°$ (opposite the CO), $+60°$ (opposite the H) or $+180°$ (opposite the NH). The most common values are near $-70°$ (Janin and Wodak, 1978). Val, Ile and Thr strongly prefer the χ_1 value with one side-chain branch opposite NH and the other opposite CO.

In α-helices the χ_1 constraints become stronger, and 90% of the side chains of Val and Ile are opposite both NH and CO. In parallel sheets, since the relationship between adjacent strands is a translation, β-branched side chains can *cup* neatly against one another. Whole strips of β-branched residues are commonly found with parallel β-sheets (Lifson and Sander, 1980). For antiparallel sheets, the relationship between adjacent strands is a twofold rotation, so that neighboring β-branched residues are either back to back, leaving unfilled space, or face to face, producing a collision, unless their backbone is adjusted. This brings about an alternation of branched and unbranched C_β along strips perpendicular to the strands.

Histidine, phenylalanine, tyrosine and tryptophan

All the amino acid residues involve the aromatic ring in their molecules, and except for His the other side chains are hydrophobic. For all the chains there is a C_β methylene group between C_α and the aromatic ring. With this single methylene group the side chain flexibility is rather restricted. Without the methylene group, however, the aromatic ring would lead to severe steric hindrance at the C_α atom, and the main chain would become too stiff.

Phe is completely hydrophobic, and has a strong preference for regular secondary structure. The polar hydroxyl group of Tyr dissociates only at high pH, because its pK value is 10.1. Under physiological conditions it is in an undissociated state, so that Tyr around neutral pH is hydrophobic, although the dipole has little effect on its hydrophobicity. Trp, the largest side chain, is slightly polar because the indole ring is heterocyclic. Therefore, Tyr and Trp are almost indifferent to inside versus outside. Histidine has a charge group whose pK is almost neutral, so that it easily gains or loses charge, depending on its surroundings. According to Rose *et al.* (1985) His is relatively buried.

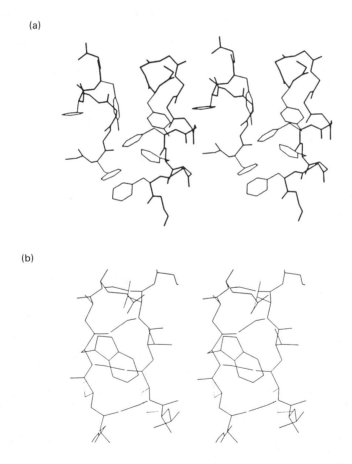

Fig. 2.29—The aromatic side chains in the protein interior. (a) An internal cluster of aromatic side chain. Almost all of them are perpendicularly stacked with one another. The structure is from cytochrome C reductase (Finzel *et al.*, 1984). (b) An aromatic side chain on a buried β-strand. The strand possesses the normally disallowed χ_1 conformation opposite the C_α hydrogen. The structure is from dihydrofolate reductase (Bolin *et al.*, 1982). (Both are cited from Richardson and Richardson, 1989.)

Phe, Tyr and Trp are major constituents of the hydrophobic core inside proteins. They pack against the peptides of the backbone, against the methyls of aliphatic side chains or the sulfurs of Met or disulfides, or against one another. This sort of packing is important for nuclear magnetic resonance (NMR) analysis of proteins (Wuthrich, 1986). The ring–ring orientations of side chains are nearly random, but frequently perpendicular stacking is found (Fig. 2.29a) (Blundell et al., 1986). The aromatic rings cannot pack well against their own local backbone in repetitive conformation. They can pack against the backbone of a neighboring β-strand (Fig. 2.29b), or against the backbone of a neighboring α-helix at moderate separation or its side chains at wide separations.

The aromatic rings are large and rigid structures with relatively few conformational degrees of freedom. In α-helices, however, the aromatic rings can stay at the best χ_1 optima: $\chi_1 = -60°$ (opposite the backbone CO) and $\chi_1 = +180°$ (opposite the NH). On loops or on the exposed part of β-sheet, there is an extremely strong preference for the χ_1 opposite the CO. For the aromatics on the inside surface of antiparallel or parallel β-sheet, which is buried on both sides (Richardson, 1981), there is a mixture of all three χ_1 conformations: opposite the NH, H and CO.

The aromatics are large enough to affect the surrounding structures, secondary and tertiary, in a specific way by their presence and local conformation: a strong influence on two-stranded antiparallel β-ribbons, encouraging both twist and curl by their interactions with backbone and with one another on the concave inner surface of a β-ribbon (Salemme, 1983).

Serine and threonine

Polar and neutral side chains are those of Cys, Ser, Thr, Asn, Gln and Tyr. The properties in forming protein structures of the residues other than Ser and Thr are given elsewhere.

The side chains of Ser and Thr are short, with an OH group. Both of them, especially Ser, are chemically reactive, so that they have important roles as catalytic site residues in serine protease and can be phosphorylated as in glycogen phosphorylase (Dixon and Webb, 1979).

The OH can be either an H-bond donor or acceptor. Ser and Thr are commonly found as the partner of H-bonds with main chain or solvent. For example, an H-bond can be formed with one of the three NHs in the first turn of an α-helix and even in the middle of helix. The latter makes a second shared H-bond to the peptide CO in the same sort of arrangement that waters often occupy next to a helical H-bond. This may exert a destabilizing effect on the α-helix, because Ser is the most unfavorable residue in the α-helix. The χ_1 angles for Ser and also Thr in α-helices show a strikingly different distribution from those of other residues. The side chains of Ser and Thr take the two χ_1 angle positions at those opposite the CO and the H. The H-bonded Ser may disturb the helix conformation.

Ser is very often located in tight turns and other non-repetitive structures, making H-bonds to neighboring backbone NH or CO groups. Since Ser is not hydrophobic and can easily interact with water, it is commonly found in the exposed location of turns and loops, and also in disordered regions. The branched β-carbon of Thr favors extended conformation, and packs well on the surface of β-sheet. In β-sheet, Thr does not make an H-bond to a peptide, but it may react indirectly through a water molecule. The extra methyl group makes Thr more hydrophobic than Ser, so that Thr occurs more often in the protein interior and less often in turns, non-repetitive loops and disordered structures, which are exposed to solvent.

Asparagine and glutamine

The side chains of Asn and Gln end in amide groups, which are similar to the peptide groups in structure. Asn, one methylene group shorter than Gln, has a geometry similar to that of a peptide, when χ_2 ($C_\beta - C_\gamma$ angle) is 180°, as shown in Fig. 2.30. In some cases the O_δ of Asn can make H-bonds as they would have been made by the O_{n-1} carbonyl group. The H-bonding in which O_δ is an H-acceptor gives a *pseudocommon* turn (Tainer *et al.*, 1982) or *Asx turn* (Rees *et al.*, 1983). An Asn or Gln side chain takes the place of the first peptide in the turn. This configuration is common, since the most frequent Asn H-bond is between O_δ and the backbone NH of residue $n + 2$ (Richardson, 1981; Baker and Hubbard, 1984). The preference of Asn for position 2 of the right turn (Chou and Fasman, 1978a, 1978b) can be explained by the characteristics of Asn described above.

Asn can depart from the main chain in another direction, while the side chain still plays the part of one or more helical residues, including an H-bond between the O_δ and the NH_{n+3}. In this way only Asn, a weak α-helix former (Chou and Fasman, 1978a, 1978b), participates in forming α-helix. Gln, on the other hand, is found in α-helix, indicating its capability for helix formation. Gln is long and flexible enough to pack on the outside of the helix rather than disrupting or competing with the backbone as does Asn. In non-repetitive loops both Asn and Gln make local side-chain-to-main-chain H-bonding.

As described above, the single methylene group by which Asn and Gln differ gives them remarkably different properties. In spite of such differences, Gln is the most

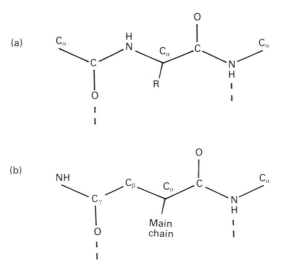

Fig. 2.30—Similarity in the geometry and H-bonding between a peptide (a) and Asn side chain with the peptide following it (b).

conservative substitution for Asn, but only where the role is amide H-bonding of a geometrically non-specific sort. Ser could be the most conservative replacement for Asn, as a helix starter, because of their position specificity as well as their preference for that position. In the first or third position of a turn, Asp could be the most conservative, while Gly is the conservative substitution in inverse turns or other left-handed conformations of Asn. In all roles involving unusual conformations, Asn is the conservative replacement for Gly. Considering such structural propensity of amino acid residues, one could replace any amino acid with another by site-directed mutagenesis.

Aspartate and glutamate

Asp and Glu are similar to the Asn and Gln pair in every respect except that they are charged at physiological pH.

Because of charged groups, Asp and Glu are generally on the protein surface and form salt bridge (H-bonds with oppositely charged groups). In the interior they are almost always making salt bridges, often with double H-bonding with Arg (Honing et al., 1986). Neighboring opposite charges on the protein surface sometimes interact, directly or through ordered water molecules. For example, the interactions occur between the side chains n and $n+3$ (or $n+4$) on an α-helix, and more often they join residues distant from one another in the sequence, and most commonly the Asp or Glu at the C-terminus is joined to the Lys at the N-terminus.

As described previously, the unique role of Asn comes from the H-bonding capability of the amide δ-oxygen rather than the δ-nitrogen. Thus, the carboxyl δ-oxygen of Asp can play similar roles in unusual H-bonding, although the ability is partly diminished by the charge. Asp can form the H-bond between its O_δ and the NH of residue $n+3$, acting as an α-helix starter. In addition, Asp is a prominent turn-forming residue, its positional preference being similar to Asn (Chou and Fasman, 1978a, 1978b). To a lesser extent of positional specificity, Glu is also a turn-forming residue.

An individual peptide has a small dipole: a partial positive charge on the N and a partial negative charge on the O. Thus, in a helix, all of the peptide dipoles are lined up, and give a partial positive charge at the N-terminus and a negative charge at the C-terminus of the helix. The helix dipole will be neutralized by the side chains at both termini and, in fact, by charged ligands (Pflugrath and Quiocho, 1985). Asp and Glu strongly prefer to the N-terminus of α-helix (in the first turn of helices), while Lys, Arg and His are more common toward the C-terminus.

Lysine and arginine

Lys and Arg are positively charged at physiological pH, due to an ε-amino group (Lys) and a guanidinium group (Arg) at the end of the long, hydrophobic side chains.

Lys and Arg locate almost exclusively on the protein surface, and they can form salt links, specifically and non-specifically, with negatively charged macromolecules such as nucleic acids.

Lysines are commonly found in disordered loops or tails. The ε-amino group of Lys has three freely rotating hydrogens, which are able to interact with water molecules. Thus, Lys is the greatest solubilizer of globular proteins. The side chain

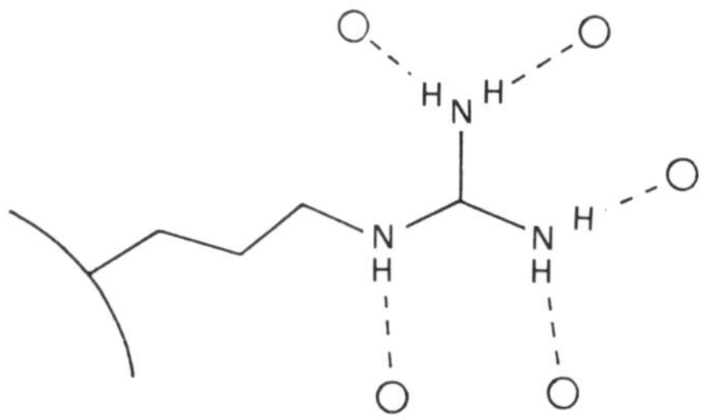

Fig. 2.31—Illustration of the guanidinium group of an Arg side chain, of which five H-atoms form the planar array of H-bonds with five O-atoms either from protein or solvent.

is indeed flexible, in spite of the considerable hydrophobicity of the side chain (Rose et al., 1985).

Arg side chains are mostly ordered among the 20 side chains, and are more buried than the Lys side chain, but not totally buried. They make extensive van der Waals interactions along the side chain, and its flat hydrophobic surface produced by curling of the side chain can be retained by replacing with Ile (Low et al., 1976).

The guanidinium groups of Arg have five H-bond donors which are held in a large, rigid, planar array. The H-bond acceptor, such as oxygen, can be arranged in optimal configuration. If the acceptor is from water, the bond angles are not optimal, and the entropy is unfavorable. Protein oxygen, on the other hand, produces conformationally satisfactory H-bonding. It is reasonable that the water molecules interact with Arg less often than the protein side chain does.

The carboxyl oxygen of Asp, Glu and C-terminus bind with Arg. The OH group of Ser, Thr or Tyr is another candidate for H-bond, and in some cases the carboxyl oxygen of Asn and Gln is found to interact with Arg. The H-bonding with Arg in crambin is shown in Fig. 2.31, and an H-bond to the OH of Thr 2 and bonds with two ordered waters are given. Replacement of Arg 17 with a Lys in crambin may have disastrous results on stability, which may partly explain the frequent replacement of Lys by Arg in thermophilic proteins (Menendez-Arias and Argos, 1989).

Proline

Pro has unique structure, as shown in Fig. 2.2. When Pro is in a peptide chain, the ring closure fixes the ϕ near $-60°$, with the two ψ around $160°$ and, less dense around $-30°$. Further, the pyrollidine ring occupies some of the space for the neighboring piece of chain, and there is no hydrogen available to H-bond for α-helix and β-sheet. Thus Pro acts as a breaker of regular secondary structure. However, partly by default Pro occurs in turns, non-repetitive structure, and at the ends of strands and helices.

Pro is not entirely prohibited in α-helices. It can be found in the first turn, and even in the central position, although it rarely occurs. Figure 2.32 shows one of the helices with an internal Pro.

Pro is also rare in the interior of β-sheet, and it is common as a *breaker* at the end of a strand, on edge strands, and in β-bulges. Pro has a strong tendency for making turns. It occurs especially in the second position of *common* and *glycine turns*,

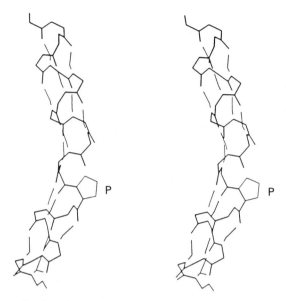

Fig. 2.32—Sterero drawing of the effect of a Pro in the middle of an α-helix. P, Pro side chain. The structure is from cytochrome C' (Finzel et al., 1985). The bend is produced, and the two H-bonds are broken. (Redrawn from Richardson and Richardson, 1989.)

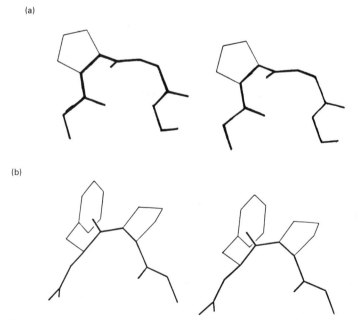

Fig. 2.33—Locations of Pro in turns. (a) Pro in position 2 of a Gly turn is preceded by *trans* peptide. (b) Pro is preceded by *cis* peptide in a type VIb turn. Both structures are from *Strep. griseus* protease (James et al., 1980). (Redrawn from Richardson and Richardson, 1989.)

both of them requiring ϕ close to $-60°$. The value of ϕ is similar to that for Pro-peptide. Thus, Pro is always exposed on the protein surface, although its chemical nature is hydrophobic. Pro is then considered to be hydrophilic.

Another peculiar property of Pro is that it occurs with a *cis* rather than *trans* conformation of peptide bond (Fig. 2.33). *Cis* peptides for other amino acids are few in number. The isomerization between *cis* and *trans* of the prolyl peptide is slow (Brandts *et al.*, 1975), so that this process may be responsible for a slow step in protein-folding kinetics.

When Pro is in permissive positions and its presence at these positions is only for some structure-specific need, it makes a positive contribution to protein stability. As seen from the structure of Pro, one degree of freedom is already frozen out in the denatured state. Thus, Pro has a smaller loss of entropy on folding than any other residue. This has been verified by substituting Pro for Ala where the structure is permitted in T4 lysozyme (Matthews *et al.*, 1987).

Cysteine

Three different types of Cys are found in globular proteins: free SH, ligand SH, and disulfide. The conformation of disulfide and its contribution to protein structure of disulfides will be described in the following.

Cysteine looks like Ser in structure, but there are many differences between them. The OH of Ser, not SH of Cys, is phosphorylated, and the SH is much more oxidizable and reactive to many reagents, e.g. heavy metals such as the mercurial compounds. The mercurials are employed as markers for X-ray crystal analysis. Because of the sensitivity of SH, Cys is relatively uncommon, and it occurs predominantly under reducing circumstances, such as in bacterial cells.

Cysteine is much poorer in H-bonding capability than Ser. Thus replacement of Cys by other amino acids has been attempted by site-directed mutagenesis (Ghosh *et al.*, 1986). Ser is hydrophilic and therefore often occurs on the outside of proteins. Cysteine, on the other hand, is buried (Rose *et al.*, 1985), and experimental results on the Cys molecule indicate that Cys is neutral or only moderately hydrophobic (Wolfenden *et al.*, 1981). The hydrophobic conditions around Cys may protect it from

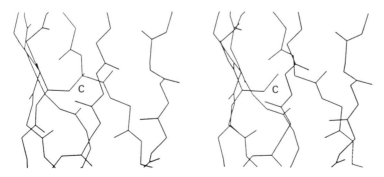

Fig. 2.34—Stereo view of a free-SH of Cys (C) in its preferred location. The Cys points inward on a β-strand of Cu, Zn superoxide dismutase (Tainer *et al.*, 1982). (Redrawn from Richardson and Richardson, 1989.)

deleterious damage; e.g. oxidative destruction. Free Cys prefers regular secondary structures to turns or coils. Its favorite location is on β-strand pointing inward and buried (Fig. 2.34).

Cysteines can serve as ligands for various metal ions and prosthetic groups, such as Fe, Zn, Cu, Fe—S cluster (in ferredoxin) and hemes. Two to four Cys, some or all of them being close together in the sequence, are clustered together to bind a group. Two liganded Cys are commonly separated by only one or two intervening residues at or near the end of a loop or on successive turns of distorted helix. Liganded cysteines are commonly located in coil regions, less so in helix, but quite rare in β-sheet.

Cystine (disulfide)

The covalent bonds which participate in constructing protein conformations are disulfides as well as peptide bonds, and the disulfides also participate in forming the primary protein structure. Moreover, the disulfides give extra stability to proteins (Anfinsen and Scheraga, 1975). Cross-linkage by disulfide bridges between pairs of Cys is formed within a single polypeptide chain and also between different chains. A given Cys combines with only one other Cys, so that the set of disulfide bonds is quite unique for a given protein (e.g. Anfinsen *et al.*, 1961; Anfinsen and Scheraga, 1975).

The two Cys which form a disulfide bond are in general located apart from each other in the sequence. As a rule the bonds form spontaneously *in vitro*, and no external factors such as enzymes are necessary (Anfinsen and Scheraga, 1975). Exceptions, however, have been reported (Givol *et al.*, 1965; Della-Corte and Parkhouse, 1973).

Two α-carbons should be separated by 4–7.5 Å, and the C_α—$C_{\alpha'}$ vector must have the right sort of relationship. The structure of a disulfide bond is given in Fig. 2.35. In Fig. 2.36 are presented the two sterically different configurations of the disulfide bond: the left-handed spiral and the right-handed hook. The former, the most common disulfide conformation (Richardson, 1981), has the antiparallel but offset C_α—$C_{\alpha'}$ vectors, while the latter, the C_α—$C_{\alpha'}$ vectors, are approximately perpendicular. The average values of χ_3 and the C_α—$C_{\alpha'}$ distance are $85 \pm 9°$ and $99 \pm 11°$, and 5.88 ± 0.49 Å and 5.07 ± 0.73 Å, respectively, for the left- and right-handed disulfides (Katz and Koshiakoff, 1986).

Subtilisins with engineered disulfides at various positions exhibit less or no enzyme activity (Mitchinson and Wells, 1989). The results may be due to the constraint or strain of the protein conformation derived by the deviation from the standard values of the above parameters.

A disulfide with a strong twist can be formed between adjacent antiparallel β-strands (Kline *et al.*, 1986). Commonly disulfides occur with a strong twist just past

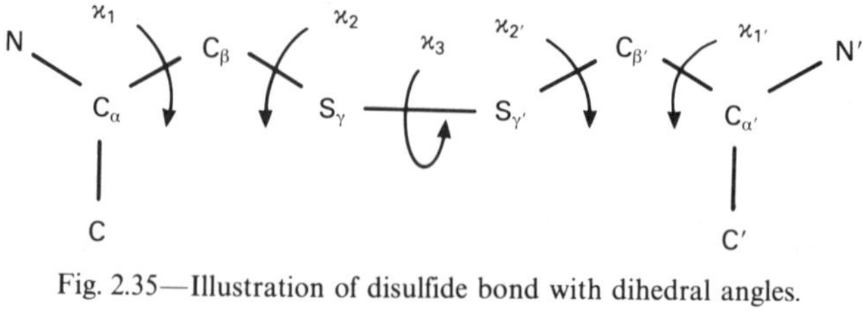

Fig. 2.35—Illustration of disulfide bond with dihedral angles.

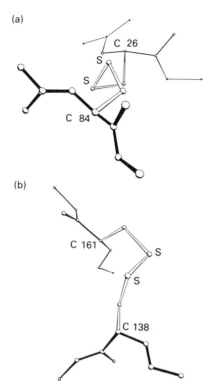

Fig. 2.36—Conformations of two common disulfide bonds: (a) left-handed spiral; (b) right-handed hook. (Cited from Richardson and Richardson, 1989.)

the ends of a β-strand pair. The disulfide is offset diagonally relative to the β-strand pairing. A disulfide cannot form in the same α-helix nor between two α-helices, without extreme distortion.

2.5 QUATERNARY STRUCTURE

In the preceding sections (2.3 and 2.4) were described the structural events for monomeric proteins composed of a single polypeptide chain. Most intracellular proteins, however, are oligomeric proteins which are made up from two or more peptide chains (subunits) of the same or different amino acid sequences. In almost all cases, oligomeric proteins can exhibit their own functions only when all the subunits are aggregated—no activity is observed with the separated subunits. One of the exceptional proteins is aspartate transcarbamylase (Jacobson and Stark, 1973). The enzyme is composed of six catalytic and six regulatory subunits, and it can be dissociated into the two kinds of subunit proteins without loss of the independent catalytic and regulatory functions.

Contacting forces

Before describing the properties of the contacting surface between subunits (or protein–protein interface), the concept of accessible surface area will be presented.

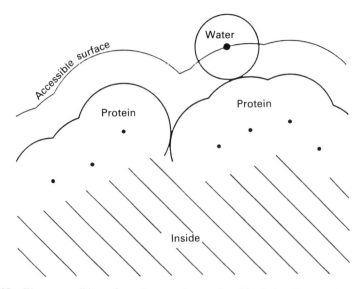

Fig. 2.37—Water-accessible surface of a protein or of a side chain. The protein surface is described by van der Waals envelopes of surface atoms indicated by dots. The protein interior is hatched. Water molecules are considered as spheres of 1.4 Å radius. The water-accessible surface is defined as the area described by the center of a water molecule that rolls over the van der Waals envelope of the protein or of the side chain. The protein or side chain surface areas in clefts are inaccessible. (Redrawn from Schulz and Schirmer, 1979.)

The accessible surface area or water-accessible surface area is the extent to which protein atoms (amino acid side chains) can form contacts with the solvent water (Chothia, 1974). For a given atom it is defined as the area over which the center of a water molecule can be placed while retaining van der Waals contact with that atom and not penetrating any other atom (Fig. 2.37). Accessible surface areas of atoms are correlated with their hydrophobicities (or hydrophobic free energies) (Chothia, 1974).

On aggregation of subunits, about 10–20% of the accessible surface area of each subunit is reduced. The average numbers of non-polar side chains located in the subunit–subunit interface of proteins are not greater than those on the surface of the aggregates of subunits (Chothia and Janin, 1975). However, a free energy term favoring association is derived from the decrease in surface area accessible to water, and the hydrophobic contribution is the major factor which stabilizes the associates. In the interface most polar side chains are H-bonded and charged groups form salt bridges. They decide the subunit–subunit association. The interface is closely packed, as in the protein interior.

Symmetry of aggregates

As in crystals, the subunits are packed in a protein so as to minimize the free energy in the associated state. Thus symmetric arrangements of subunits are energetically favored rather than asymmetric arrangements. Most commonly observed symmetries— point symmetries—are represented in Fig. 2.38.

2.5 Quaternary Structure

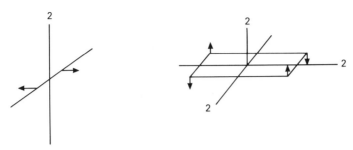

Fig. 2.38—Point symmetries. Representations of point groups 2 (C_2), and 222 (D_2). The numbers denote twofold, threefold, etc. (For further details see Hammermesh, 1962.)

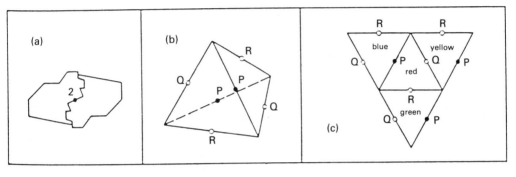

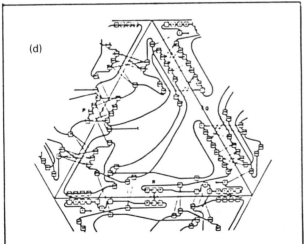

Fig. 2.39—Representation of intersubunit contacts of oligomeric enzymes (dehydrogenases). (a) A symmetric dimer, e.g. alcohol dehydrogenase with only one type of interface. The contacting surfaces are complementary. (b) Tetramer with 222 symmetry represented by a tetrahedron. Each subunit is represented by a triangle and each intersubunit contact by an edge. The three twofold axes are indicated as P, Q and R. Among the three different contact types only two have to be formed to keep all subunits from falling apart. (c) Planar representation of the tetrahedron and of subunit contacts. (d) Use of the planar representation for explicit specification of the atomic contacts between subunits of lactate dehydrogenase (Liljas and Rossmann, 1974). (Redrawn from Schulz and Schirmer, 1979.)

As shown in Fig. 2.38, soluble malate and alcohol dehydrogenases are dimers of identical subunits with point group 2. The dimers have only one type of contact around the twofold axis. Lactate and glyceraldehyde 3-phosphate dehydrogenases are tetramers of identical subunits with point group 222. The tetramers have three types of contacts: one across each twofold axis.

The tetramer contacts are conveniently represented by the tetrahedron or by the two-dimensional plot (Fig. 2.39). In the representations, the subunits are marked using a color code, and the twofold axes are denoted P, Q and R. The *red* subunit is the reference base, and it forms contacts with *blue*, *yellow* and *green* across the twofold axes, P, Q and R, respectively. The two-dimensional plot is used for indicating the interactions between residues of the *red* subunit with residues on the other subunits.

2.6 SIDE CHAINS AND WATER

Protein folding is a self-assembly capacity of polypeptide chains, and many aspects of protein configuration derive from the properties of the particular sequence of amino acids that make up the polypeptide chain. These properties include the characteristics of both the individual side chains and the polypeptide backbone (Epstein *et al.*, 1963). The interactions which occur between the polypeptide backbone, between the side chains and between the backbone and side chains to fold the polypeptide chain are H-bonding, electrostatic and hydrophobic interactions and dispersion forces. As described in section 2.3.1.5, the folding is determined by the amino acid sequence, and is driven by the entropy of removing hydrophobic side chains from contact with water (Kauzmann, 1959).

Hydrophobicity

Hydrophobic interactions are now considered to play an important role in protein folding as well as—although maybe not exclusively—in protein stabilization. The hydrophobic scales of amino acid side chains will be briefly described.

The hydrophobicity of side chains was first quantitatively evaluated by estimating the free energy of the transfer of amino acids from water to an organic solvent, such as ethanol, as a model of the protein interior (Tanford, 1963; Nozaki and Tanford, 1971). Later, considering the distribution probability of amino acids in real proteins, i.e. whether on the surface or buried partially or entirely in the interior, several hydrophobic scales have been proposed (Chothia and Janin, 1975; Janin, 1979; Ponnuswamy *et al.*, 1980; Wolfenden *et al.*, 1981; Kyte and Doolittle, 1982; Rose *et al.*, 1985). No single correct answer to the quantitative hydrophobicity has yet been presented. For quantitative estimation of the contribution of hydrophobicities of engineered amino acids to protein stability (see Chapters 3 and 7), however, the hydrophobic scales of Tanford are still employed (for more details see Chapter 3).

The hydrophobic effects principally participate in the hydrophobic side chains in the protein interior and the hydrophilic chains on the surface. However, some of the charged and ionized side chains can be inside the surface, the buried chains interacting with each other, and large hydrophobic chains are often found on the surface.

Ordered structures and water

A hydrophobic periodicity of about 3.6 residues is found in almost all α-helices, and many hydrophobic side chains are arranged on one half of the helices and the hydrophilic chains on the other half (Schiffer and Edmundson, 1967). Thus the great majority of α-helices in globular proteins are at the surface, approximately half exposed to water and half buried. Bacteriorhodopsin, completely embedded in the membranes of *Halobacterium halobium*, consists of seven subunits composed of α-helix (Henderson and Unwin, 1975; Blaurock, 1975). Similarly to globular proteins, these helices in the membrane-bound protein have their hydrophobic sides favoring the non-polar lipid moiety, whereas more hydrophilic proteins are in the interior of the molecule (Engelman and Zaccai, 1980). The 3.6-residue periodicity is discontinued at both N- and C-termini, which are hydrophilic.

On antiparallel β-structure the β-strands alternate side-chain direction toward the opposite sides of the β-sheet; the hydrophobic and hydrophilic side chains are alternately arranged, as shown in Fig. 2.13. Thus, one side of the strand is buried and the other is exposed to water. Parallel β-strands, on the other hand, are typically buried on both sides, so that the regions near the center of a sheet are hydrophobic, but those around the sheet end hydrophilic. The side-chain-to-main-chain H-bonding capabilities of hydrophilic residues can form the turns and non-repetitive loops (section 2.3.4). Thus these structures are hydrophilic and almost always exposed on the surface (Leszczynski and Rose, 1986). Completely disordered structure is the most hydrophilic of all, because the structure lacks large hydrophobic side chains but has an abundance of residues which interact well with water.

Solvent effects

Water is statistically in a more ordered state when next to hydrophobic side chains than in bulk solvent without any enthalpic compensation provided by H-bonding such as with hydrophilic side chains. It is therefore more favorable for hydrophobic groups to interact with each other and waters with waters than for hydrophilics to interact with water. As mentioned before, this is the driving force for folding proteins in water.

Another type of solvent effect on protein structures is the specific side chain or peptide groups in particular positions and conformations. As stated in section 2.4.3, Lys interacts with waters strongly but in a highly mobile and disordered mode, while Arg interacts with waters and peptide carbonyls in nearly planar array. Asp, Glu, Asn and Gln can make lesser but similar H-bonding arrays, some of them with water. Hydroxyl groups interact very specifically with water, but visible electron-density peaks for ordered water are often unobserved, because of the rotationally disordered conformation of the OH groups. Charged groups on the surface may not form direct salt links with their opposite counterparts, but prefer to make a statistical network of indirect salt links through shared water molecules.

Association of subunits or macromolecular complexes is driven by the change in surface area buried from water (Chothia, 1976). Although these buried surfaces are often tightly fitted as in the protein interior, a single layer of ordered water is observed in many domain interfaces and some of the more labile subunit contacts (Richardson and Richardson, 1987). In these less hydrophobic interfaces there are relatively few direct contacts, and many H-bonds are bridged through the waters.

2.7 PREDICTION OF PROTEIN STRUCTURES

To understand the mechanism of folding and the biological function of proteins, it is necessary to have information of their three-dimensional conformation. X-ray crystallographic analysis has been successfully employed for determining the special structure of proteins. As is well known, however, it is no easy task to obtain a large single crystal of protein, and in spite of remarkable advances in computer technique it is still laborious and time consuming to determine protein structure at the atomic level. Protein engineering, which has become a promising tool for elucidating the mechanism of protein stability (see Chapter 3) and for stabilizing proteins (Chapter 6), also strongly necessitates the three-dimensional structure of protein to be engineered.

Since the first demonstration of reversible unfolding of ribonuclease (Anfinsen *et al.*, 1961), the native conformation of a protein is coded in its amino acid sequence. Thus, many efforts have been made to predict protein structure from the sequence data.

2.7.1 Secondary structure

So far many approaches for the prediction of the secondary structure of proteins have been proposed: Kotelchuck and Scheraga (1968, 1969), Lewis *et al.* (1970), Ptitsyn and Finkelstein (1970a, 1970b), Toitskii and Zav'yalov (1972), Kabat and Wu (1973a, 1973b), Chou and Fasman (1974a, 1974b, 1978a, 1978b), Burgess *et al.* (1974), Lim (1974a, 1974b), Robson and Suzuki (1976), Garnier *et al.* (1978), Nagano (1977), McLachlan (1977), Barkovsky and Bandarin (1979) and Barkovsky (1982). The methods so far reported can be categorized into two classes. The empirical statistical methods of, for example, Chou and Fasman (1974a, 1974b) and of Garnier *et al.* (1978) use parameters obtained from the analysis of known sequences and structures. The second method of Lim (1974a, 1974b) is based on a stereochemical theory of globular proteins. The methods most frequently used have been the empirical approaches of Chou and Fasman and of Robson, and the stereochemical approach of Lim.

The Chou–Fasman method is one of the most frequently used for secondary structure prediction. This is because of its simplicity and its reasonably high degree of accuracy. In the following, then, the outline of the procedure and its applicability to predict secondary structure are briefly described.

The number of occurrences of a given amino acid in the α-helix, β-sheet and coil were calculated from the X-ray crystal structures of 29 proteins containing 4741 amino acid residues (Chou and Fasman, 1978a). Considering the relative frequency of a given amino acid in a protein, its occurrence in a given type of secondary structure, and the fraction of residues occurring in that type of structure, the conformational parameters for each amino acid were calculated from the results (Table 2.6).

An abbreviated set of rules for predicting protein secondary structure using these parameters follows (Fasman, 1985):

(1) A cluster of four helical residues (H_α or h_α) out of six along the sequence will initiate a helix. The helical segment is extended in both directions until sets of

Table 2.6—Conformational parameters for α-helical, β-sheet and β-turn residues in 29 proteins. P_α, P_β, and P_i are conformational parameters of helix, β-sheet and β-turn, respectively. H_α or H_β, strong α- or β-former; h_α or h_β, α- or β-former; I_α or I_β, weak α- or β-former; i_α or i_β, α- or β-indifferent; b_α or b_β, α- or β-breaker; B_α or B_β, strong α- or β-breaker. f_i, f_{i+1}, f_{i+2}, and f_{i+3} are bend frequencies in the four positions of the β-turn (Redrawn from Chou and Fasman, 1978b)

	P_α			P_β			P_i		f_i		f_{i+1}		f_{i+2}		f_{i+3}
Glu	1.51		Val	1.70	}	Asn	1.56	Asn	0.161	Pro	0.301	Asn	0.191	Trp	0.167
Met	1.45	} H_α	Ile	1.60	} H_β	Gly	1.56	Cys	0.149	Ser	0.139	Gly	0.190	Gly	0.152
Ala	1.42		Tyr	1.47		Pro	1.52	Asp	0.147	Lys	0.115	Asp	0.179	Cys	0.128
Leu	1.21		Phe	1.38		Asp	1.46	His	0.140	Asp	0.110	Ser	0.125	Tyr	0.125
Lys	1.16		Trp	1.37		Ser	1.43	Ser	0.120	Thr	0.108	Cys	0.117	Ser	0.106
Phe	1.13		Leu	1.30		Cys	1.19	Pro	0.102	Arg	0.106	Tyr	0.114	Gln	0.098
Gln	1.11		Cys	1.19	} h_β	Tyr	1.14	Gly	0.102	Gln	0.098	Arg	0.099	Lys	0.095
Trp	1.08	} h_α	Thr	1.19		Lys	1.01	Thr	0.086	Gly	0.085	His	0.093	Asn	0.091
Ile	1.08		Gln	1.10		Gln	0.98	Tyr	0.082	Asn	0.083	Glu	0.077	Arg	0.085
Val	1.06		Met	1.05		Thr	0.96	Trp	0.077	Met	0.082	Lys	0.072	Asp	0.081
Asp	1.01	} I_α	Arg	0.93		Trp	0.96	Gln	0.074	Ala	0.076	Thr	0.065	Thr	0.079
His	1.00		Asn	0.89	} i_β	Arg	0.95	Arg	0.070	Tyr	0.065	Phe	0.065	Leu	0.070
Arg	0.98		His	0.87		Met	0.95	Met	0.068	Glu	0.060	Trp	0.064	Pro	0.068
Thr	0.83	} i_α	Ala	0.83		Val	0.74	Val	0.062	Cys	0.053	Gln	0.037	Phe	0.065
Ser	0.77		Ser	0.75	} b_β	Leu	0.66	Leu	0.061	Val	0.048	Leu	0.036	Glu	0.064
Cys	0.70		Gly	0.75		Ala	0.60	Ala	0.060	His	0.047	Ala	0.035	Ala	0.058
Tyr	0.69	} b_α	Lys	0.74		Phe	0.60	Phe	0.059	Phe	0.041	Pro	0.034	Ile	0.056
Asn	0.67		Pro	0.55		Glu	0.59	Glu	0.056	Ile	0.034	Val	0.028	Met	0.055
Pro	0.57	} B_α	Asp	0.54	} B_β	Lys	0.50	Lys	0.055	Leu	0.025	Met	0.014	His	0.054
Gly	0.57		Glu	0.37		Ile	0.47	Ile	0.043	Trp	0.013	Ile	0.013	Val	0.053

tetrapeptide α-helix breaker (the sum of $\langle P_\alpha \rangle$ is below 1.00) are reached. Pro cannot occur in the inner helix or at the C-terminal helical end but can occur within the last three residues at the N-terminal end. The inner helix is defined as that omitting the three helical end residues at both the amino and carboxy terminal ends. Any segment that is at least six residues long with $\langle P_\alpha \rangle > 1.03$ and $\langle P_\beta \rangle > \langle P_\alpha \rangle$ is predicted as helical.

(2) A cluster of three β-formers or a cluster of three β-formers out of five residues along the sequence will initiate a β-sheet. The β-sheet propagates in both directions until terminated by a set of tetrapeptide breakers ($\langle P_\beta \rangle < 1.00$). Any segment with $\langle P_\beta \rangle > 1.05$ as well as $\langle P_\beta \rangle > \langle P_\alpha \rangle$ is predicted as β-sheet.

(3) The probability of a bend at residue i is calculated from $P_t = f_i \times f_{i+1} \times f_{i+2} \times f_{i+3}$. Tetrapeptides with $P_t > 1.00$ and $\langle P_\alpha \rangle < \langle P_t \rangle > \langle P_\beta \rangle$ are predicted as β-turn.

(4) Any segment containing overlapping α- and β-regions is helical if $\langle P_\alpha \rangle > \langle P_\beta \rangle$ or β-sheet if $\langle P_\beta \rangle > \langle P_\alpha \rangle$.

Using these parameters and rules, one can find nucleation sites within the sequence and extend them until a stretch of amino acids is encountered that is not disposed to occur in that type of structure or until a stretch is encountered that has a greater disposition for another type of structure. At this point, the structure is terminated. This process is repeated throughout the sequence until the entire sequence is predicted. The arithmetic calculations can be performed with a computer program, but the data reduction should be performed by hand. A knowledge of protein structures may facilitate the procedure.

Table 2.7 shows the secondary structures of staphylococcal nuclease predicted by the Chou–Fasman method and determined by X-ray analysis (Prevelige and Fasman, 1989). As shown in the table, a total of 44 amino acid residues were either overestimated or missed. However, all the secondary structure regions were correctly predicted, and a total of 71% of the residues were predicted. According to the procedure about 70–80% of predictive accuracy is obtained (Prevelige and Fasman, 1989).

Table 2.7—Prediction of the regions containing α-helix and β-sheet structures in staphylococcal nuclease (from Prevelige and Fasman, 1989)

Predicted	X-ray	Residues overpredicted	Residues missed
7–18 α	12–19 α	5	1
23–26 β	21–27 β	0	3
31–30 β	30–36 β	4	1
59–76 α	54–67 α	10	8
97–104 α	99–106 α	2	2
120–140 α	122–134 α	8	0
Total		29	15

2.7.2 Tertiary structure

There are three approaches now used for the prediction of the tertiary structures of proteins (Ponder and Richards, 1987). (1) The tertiary structure of a polypeptide is constructed from its amino acid sequence by referring to the three-dimensional conformations of proteins of which sequences are homologous with the sample. (2) The secondary structural units are predicted, and then the units are assembled into a compact structure, based on the information of protein structure. The two methods strongly depend on the database of X-ray crystal structures. The first procedure uses the rules of structures by implication, but the second one uses them quantitatively. (3) The third one is based on the use of empirical energy functions to derive the tertiary structure of minimum potential energy, relying on quantitative calculations throughout the approach.

The second method, the combinational approach, has recently been developed (Cohen *et al.*, 1979, 1980, 1981, 1983; Cohen and Sternberg, 1980; Edwards *et al.*, 1987). The procedure consists of three stages: (1) the prediction of the secondary structures, now with up to 80% accuracy; (2) the packing of the predicted α-helices and β-strands into an approximate native fold (Ptitsyn and Rashin, 1975; Cohen *et al.*, 1979, 1980, 1981, 1982; Efimov, 1979; Lifson and Sander, 1980); (3) the use of simplified energy calculations (Levitt, 1976; Kuntz *et al.*, 1976; Robson and Osguthrope, 1979) to refine the above folded conformation into the native protein. According to Hagler and Honig (1978) and Cohen and Sternberg (1980) the combinational approach has many advantages.

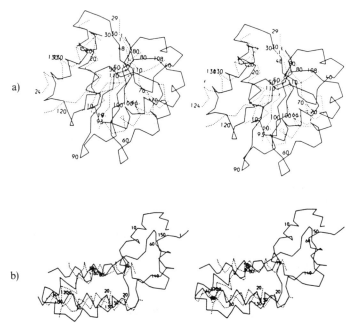

Fig. 2.40—The X-ray crystal and predicted structures of flavodoxin (a) and tobacco mosaic virus coat protein (b). α-Carbon diagrams of the crystal structures represented by continuous lines are superimposed with the predicted structures indicated by dotted lines. (Redrawn from Cohen and Kuntz, 1989.)

In the combinational approach, attention is mainly focussed on the docking of secondary structures into a native-like three-dimensional structure. Among the possible structures generated by packing all combinations of the α-helices and β-sheets, structures that violate the steric and connectivity constraints are eliminated. In the case of myoglobin (Richmond and Richards, 1978; Cohen et al., 1983; Cohen and Sternberg, 1980), over 10^8 trial structures were generated by docking together hydrophobic patches on the surface of the α-helices. Among them, only 20 folded structures satisfied the stereochemical rules. The examinations of distance constraints around heme binding site further reduced the number of allowed structures to two, and the relative arrangement of α-helices in one of them was found to closely resemble that in the native protein.

There have been many reports on the improved or advanced prediction for packing the secondary structures. For example, the supersecondary structure (Rao and Rossman, 1973) was predicted in protein sequence by Taylor and Thornton (1984). Vonderviszt et al. (1986) developed an approach to domain border prediction only from the amino acid sequence. Billeter et al. (1987) devised a new method for optimizing the constraints when two molecules are docked. This method was explored as a tool for determining sterically acceptable interactions between two molecules, i.e. the constraints on intermolecular distances.

The combinational approach has been applied to a large number of proteins of three classed (α/α, β/β and α/β), and some stereo diagrams comparing the predicted and X-ray crystal structures are shown in Fig. 2.40 (Cohen and Kuntz, 1989).

REFERENCES

Almassy, R. J., Fontecilla-Camps, J. C., Suddath, F. L. and Bugg, C. E. (1983) Structure of variant-3 scorpion neurotoxin from *Centruroides sculpuratus* Ewing at 1.8 Å resolution. *J. Mol. Biol.*, **170**, 497–527.

Anderson, A. and Hermans, J. (1988) Microfolding: conformational probability map for the alanine dipeptide in water from molecular dynamics simulations. *Proteins Struct. Funct. Genet.*, **3**, 262–265.

Anfinsen, C. B. and Scheraga, J.-P. (1975) Experimental and theoretical aspects of protein folding. *Adv. Prot. Chem.*, **29**, 205–300.

Anfinsen, C. B., Huber, E., Sela, M. and White, F. H. (1961) The kinetics of formation of native ribonuclease during oxidation of reduced polypeptide chain. *Proc. Natl. Acad. Sci. USA*, **47**, 1309–1314.

Argos, P., Rossmann, M. and Johnson, J. (1977) A four-helical super-secondary structure. *Biochem. Biophys. Res. Commun.*, **75**, 83–86.

Baker, E. and Dodson, E. J. (1980) Crystallographic refinement of the structure of actinidin at 1.7 Å resolution by fast Fourier least squares methods. *Acta Crystallogr., Section A*, **36**, 559–572.

References

Baker, E. and Hubbard, R. (1984) Hydrogen bonding in globular proteins. *Prog. Biophys. Mol. Biol.*, **44**, 97–179.

Banner, D., Bloomer, A. C., Petsko, G. A., Phillips, D. C., Pogson, C., Wilson, I., Corran, P. H., Furth, A. J., Milman, J. D., Offord, R. E., Priddle, J. D. and Waley, S. G. (1975) Structure of chicken muscle triose phosphate isomerase determined crystallographically at 2.5 Å resolution using amino acid sequence data. *Nature*, **255**, 609–614.

Banner, D., Kokkinidis, M. and Tsernoglou, D. (1987) The structure of the ColE1 ROP protein at 1.7 Å resolution. *J. Mol. Biol.*, **196**, 657–675.

Barkovsky, E. V. (1982) Prediction of the secondary structure of globular proteins by their amino acid sequence. *Acta Biol. Med. Germ.*, **41**, 751–758.

Barkovsky, E. V. and Bandarin, V. A. (1979) Secondary structure prediction of globular proteins from their amino acid sequence. *Bioorg. Khim.*, **5**, 24–34.

Billeter, M., Havel, T. F. and Wuthrich, K. (1987) The ellipsoid algorithm as a method for the determination of polypeptide conformations from experimental distance constraints and energy minimization. *J. Comput. Chem.*, **8**, 132–141.

Birktoft, J. J. and Blow, D. M. (1972) Structure of crystalline α-chymotrypsin. *J. Mol. Biol.*, **68**, 187–240.

Blake, C. C. F., Geisow, M. F., Oatley, S. J., Rerat, B. and Rerat, C. (1978) Structure of prealbumin: secondary, tertiary and quaternary interactions determined by Fourier refinement at 1.8 Å. *J. Mol. Biol.*, **121**, 339–356.

Blaurock, A. E. (1975) Bacteriorhodopsin: a trans-membrane pump containing α-helix. *J. Mol. Biol.*, **93**, 139–158.

Blundell, T., Singh, J., Thornton, J., Burley, S. and Petsko, G. (1986) Aromatic interactions. *Science*, **234**, 1005.

Bolin, J. T., Filman, D. J., Matthews, D. A., Hamlin, R. C. and Kraut, J. (1982) Crystal structure of *Escherichia coli* and *Lactobacillus casei* dihyrofolate reductase refined at 1.7 Å resolution. I. General features and binding of methotrexate. *J. Biol. Chem.*, **257**, 13650–13662.

Bondi, A. (1964) Van der Waals volumes and radii. *J. Phys. Chem.*, **68**, 441–451.

Brandts, J., Halvorson, H. and Brennan, M. (1975) Consideration of the possibility that the slow step in protein denaturation reactions is due to *cis–trans* isomerism of proline residues. *Biochemistry*, **14**, 4953–4963.

Brant, D. A., Miller, W. G. and Flory, P. J. (1967) Conformational energy estimate for statistically coiling polypeptide chain. *J. Mol. Biol.*, **23**, 47–65.

Burgess, A. W., Ponnuswamy, P. K. and Scheraga, H. A. (1974) Analysis of conformations of amino acid residues and prediction of backbone topography in proteins. *Israel J. Chem.*, **12**, 239–286.

Caspar, D. L. D., Cohen, C. and Longley, W. (1969) Tropomyosin: crystal structure, polymorphism and molecular interactions. *J. Mol. Biol.*, **41**, 87–107.

Chothia, C. (1973) Conformation of twisted β-pleated sheets in proteins. *J. Mol. Biol.*, **75**, 295–302.

Chothia, C. (1974) Hydrophobic bonding and accessible surface area in proteins. *Nature (London)*, **248**, 338–339.

Chothia, C. (1976) The nature of the accessible and buried surfaces in proteins. *J. Mol. Biol.*, **105**, 1–14.

Chothia, C. (1983) Coiling of β-pleated sheets. *J. Mol. Biol.*, **163**, 107–117.

Chothia, C. and Janin, J. (1975) Principles of protein–protein recognition. *Nature (London)*, **256**, 705–708.

Chothia, C. and Janin, J. (1981) Relative orientation of close-packed β-pleated sheets in proteins. *Proc. Natl. Acad. Sci. USA*, **78**, 4146–4150.

Chothia, C. and Janin, J. (1982) Orthogonal packing of β-sheets in proteins. *Biochemistry*, **21**, 3955–3965.

Chothia, C., Levitt, M. and Richardson, D. C. (1977) Structure of proteins: packing of α-helices and pleated sheets. *Proc. Natl. Acad. Sci. USA*, **74**, 4130–4134.

Chou, P. Y. and Fasman, G. D. (1974a) Conformational parameters for amino acids in helical, β-sheet, and random coil regions calculated from proteins. *Biochemistry*, **13**, 211–222.

Chou, P. Y. and Fasman, G. D. (1974b) Prediction of protein conformation. *Biochemistry*, **13**, 222–244.

Chou, P. Y. and Fasman, G. D. (1978a) Prediction of the secondary structure of proteins from their amino acid sequence. *Adv. Enzymol.*, **47**, 45–148.

Chou, P. Y. and Fasman, G. D. (1978b) Empirical predictions of protein conformation. *Annu. Rev. Biochem.*, **47**, 251–276.

Cohen, F. E. and Kuntz, I. D. (1989) Tertiary structure prediction. In *Prediction of protein structure and the principles of protein conformation* (Fasman, G. D., ed.), pp. 647–706. Academic Press, New York.

Cohen, F. E. and Sternberg, M. J. E. (1980) The use of chemically derived distant constraints in the predictions of protein structure with myoglobin as an example. *J. Mol. Biol.*, **137**, 9–22.

Cohen, F. E., Richmond, J. T. and Richards, F. M. J. (1979) Protein folding: evaluation of some simple rules for the assembly of helices into tertiary structure with myoglobin as an example. *J. Mol. Biol.*, **132**, 275–288.

Cohen, F. E., Sternberg, M. J. E. and Taylor, W. R. (1980) Analysis and prediction of protein β-sheet structures by a combinated approach. *Nature (London)*, **285**, 378–382.

Cohen, F. E., Sternberg, M. J. E. and Taylor, W. R. (1981) The analysis of the tertiary structure of protein sandwiches. *J. Mol. Biol.*, **148**, 253–272.

Cohen, F. E., Sternberg, M. J. E. and Taylor, W. R. (1982) The analysis and prediction of the tertiary structure of globular proteins involving the packing of α-helices against a β-sheet: a combinational approach. *J. Mol. Biol.*, **156**, 821–862.

Cohen, F. E., Abarbanel., R. M., Kuntz, I. D. and Fletterick, R. J. (1983) Secondary structure assignment for α/β proteins by a combinational approach. *Biochemistry*, **22**, 4894–4904.

Colman, P. M., Freeman, H. C., Guss, J. M., Murata, M., Norris, V. A., Ramshaw, J. A. and Venkatappa, M. P. (1978) X-ray crystal structure analysis of plastcyanin at 2.7 Å resolution. *Nature (London)*, **272**, 319–324.

Cook, W., Einspahr, H., Trapane, T., Urry, D. and Bugg, C. (1980) The crystal structure and conformation of the cyclic trimer of a repeat pentapeptide of elastin: Cyclo L Val–L Pro–Gly–L Val–Gly. *J. Am. Chem. Soc.*, **102**, 5502–5505.

Cotton, F. A., Hazen, E. E., Jr and Legg, M. J. (1979) Staphylococcal nuclease: proposed mechanism of action of based on structure of enzyme–thymidine 3′,5′-bisphosphate–calcium ion complex at 1.5 Å resolution. *Proc. Natl. Acad. Sci. USA*, **76**, 2251–2255.

Creighton, T. E. (1983) *Proteins*. Freeman, New York.

Della Corte, E. and Parkhouse, K. I. (1973) Biosynthesis of immunoglobulin A (IgA) and immunoglobulin M (IgM). Requirement for J-chain and a disulfide-exchanging enzyme for polymerization. *Biochem. J.*, **136**, 597–606.

Dixon, M. and Webb, E. C. (1979) *Enzymes*, 3rd. edn. Longman, London.

Dodson, E. J., Dodson, G. G., Hodgkin, D. C. and Reynolds, C. D. (1979) Structural relationship in the two-zinc insulin hexamer. *Can. J. Biochem.*, **57**, 469–479.

Edwards, M., Sternberg, M. and Thornton, J. (1987) Structural and sequence patterns in the loops of βαβ units. *Protein Eng.*, **1**, 173–181.

Efinov, A. (1979) Packing of alpha-helices in globular proteins: layer-structure of globin hydrophobic cores. *J. Mol. Biol.*, **134**, 23–40.

Efinov, A. (1982) Role of constrictions in formation of protein structures containing four helical regions. *Mol. Biol. (Mosk.)*, **16**, 271–281.

Engelman, D. M. and Zaccai, G. (1980) Bacteriorhodopsin in an inside-out protein. *Proc. Natl. Acad. Sci. USA*, **77**, 5894–5898.

Epstein, C., Goldberg, R. and Anfinsen, C. (1963) The genetic control of tertiary protein structure: studies with model system. *Cold Spring Harbor Symp. Quant. Biol.*, **28**, 439–449.

Fasman, G. D. (1985) A critique of the utility of the prediction of protein secondary structure. *J. Biosci.*, **8**, 15–23.

Fermi, G., Perutz, M. F., Shaana, B. and Fourme, R. (1984) The crystal structure of human deoxyhaemoglobin at 1.74 Å resolution. *J. Mol. Biol.*, **175**, 159–174.

Finzel, B. C., Poulos, T. L. and Kraut, J. (1984) Crystal structure of yeast cytochrome C peroxidase refined at 1.7 Å resolution. *J. Biol. Chem.*, **259**, 13027–13036.

Finzel, B. C., Weber, P. C., Hardman, K. D. and Salemme, F. R. (1985) Structure of ferricytochrome C′ from *Rhodospirillum molischianum* at 1.67 Å resolution. *J. Mol. Biol.*, **186**, 627–643.

Furey, W., Jr, Wang, B. C., Yoo, C. S. and Sax, M. (1983) Structure of a novel Bence-Jones protein (Rhe) fragment at 1.6 Å resolution. *J. Mol. Biol.*, **167**, 661–692.

Garnier, J., Osguthrope, D. J. and Robson, B. (1978) Analysis of the accuracy and implications of simple methods for predicting the secondary structure of globular proteins. *J. Mol. Biol.*, **120**, 97–120.

Ghosh, S., Bock, S., Rokita, S. and Kaiser, E. (1986) Modification of the active site of alkaline phosphatase by site-directed mutagenesis. *Science*, **231**, 145–148.

Givol, D., DeLorenzo, F., Goldberger, R. F. and Anfinsen, C. B. (1965) Disulfide interchange and the three-dimensional structure of proteins. *Proc. Natl. Acad. Sci. USA*, **53**, 676–684.

Hagler, A. T. and Konig, B. (1978) On the formation of protein tertiary structure on a computer. *Proc. Natl. Acad. Sci. USA*, **75**, 554–558.

Hammermesh, M. (1962) *Group Theory*. Addison-Wesley, London.

Hecht, M. H., Sturtevant, J. M. and Sauer, R. T. (1986) Stabilization of lambda repressor against thermal denaturation by site-directed Gly → Ala changes in α-helix. *Proteins Struct. Funct. Genet.*, **1**, 43–46.

Henderson, R. and Unwin, P. N. T. (1975) Three-dimensional model of purple membrane obtained by electron microscopy. *Nature (London)*, **257**, 28–32.

Hendrickson, W. A. and Love, W. E. (1971) Structure of lamprey haemoglobin. *Nature (London)*, **232**, 197–203.

Hendrickson, W. A., Love, W. E. and Karle, J. (1973) Crystal structure analysis of sea lamprey hemoglobin at 2 Å resolution. *J. Mol. Biol.*, **74**, 331–361.

Hol, W., van Duijnen, P. and Berendsen, H. (1978) The α-helix dipole and the properties of proteins. *Nature (London)*, **273**, 443–446.

Honing, B., Hubbell, W. and Flewelling, R. (1986) Electrostatic interactions in membranes and proteins. *Annu. Rev. Biophys. Bioeng.*, **15**, 163–193.

Horwich, A. L., Neupert, W. and Hartl, F.-U. (1990) Protein-catalyzed protein folding. *Trends Biotech.*, **8**, 126–131.

Ikai, A. (1980) Thermostability and aliphatic index of globular proteins. *J. Biochem. (Tokyo)*, **88**, 1895–1898.

Irwin, M. J., Nyborg, J., Reid, B. R. and Blow, D. M. (1976) The crystal structure of tyrosyl-transfer RNA synthetase at 2.7 Å resolution. *J. Mol. Biol.*, **105**, 577–586.

IUPAC–IUB Commission on Biochemical Nomenclature (1970) Abbreviations and symbols for the description of the conformation of polypeptide chains. *J. Biol. Chem.*, **245**, 6489–6497.

Jacobson, G. R. and Stark, G. R. (1973) Aspartate transcarbamylases. *Enzymes*, **9**, 226–309.

James, M. N., Sielecki, A. R., Brayer, G. D., Delbaere, L. T. and Bauer, C.-A. (1980) Structures of product and inhibitor complexes of *Streptomyces griseus* protease A at 1.8 Å resolution: a model for serine protease catalysis. *J. Mol. Biol.*, **144**, 43–88.

Janin, J. (1979) Surface and inside volumes in globular proteins. *Nature (London)*, **277**, 491–492.

Janin, J. and Wodak, S. (1978) Conformation of amino acid side chains in proteins. *J. Mol. Biol.*, **125**, 357–386.

Kabat, E. A. and Wu, T. T. (1973a) The influence of nearest-neighbor amino acids on the conformation of the middle amino acid in proteins: comparison of predicted and experimental determination of β-sheets in concanavalin A. *Proc. Natl. Acad. Sci. USA*, **70**, 1473–1477.

Kabat, E. A. and Wu, T. T. (1973b) The influence of nearest-neighboring amino acid residues on aspects of secondary structure of proteins: attempts to locate α-helices and β-sheets. *Biopolymers*, **12**, 751–774.

Katz, B. A. and Koshiakoff, A. (1986) The crystallographically determined structures of atypical strained disulfides engineered into subtilisin. *J. Biol. Chem.*, **261**, 15480–15485.

Kauzmann, W. (1959) Some factors in the interpretation of protein denaturation. *Adv. Prot. Chem.*, **14**, 1–63.

Kline, A., Braun, W. and Wuthrich, K. (1986) Studies by ^{1}H nuclear magnetic resonance and distance geometry of the solution conformation of the α-amylase inhibitor Tendamistat. *J. Mol. Biol.*, **189**, 377–382.

Kretsinger, R. H. and Nockolds, C. E. (1973) Carp muscle calcium-binding protein. II. Structure determination and general description. *J. Biol. Chem.*, **248**, 3313–3926.

Kotelchuck, D. and Scheraga, H. A. (1968) The influence of short-range interactions on protein conformation. I. Side-chain–backbone interactions with a single peptide unit. *Proc. Natl. Acad. Sci. USA*, **61**, 1163–1170.

Kotelchuck, D. and Scheraga, H. A. (1969) The influence of short-range interactions on protein conformation. II. A model for predicting the α-helical regions of proteins. *Proc. Natl. Acad. Sci. USA*, **62**, 14–21.

Kuntz, I. D., Crippen, G. M., Kollman, P. A. and Kimelman, D. (1976) Calculation of protein tertiary structure. *J. Mol. Biol.*, **106**, 983–994.

Kyte, J. and Doolittle, R. (1982) A simple method for displaying the hydrophobic character of a protein. *J. Mol. Biol.*, **157**, 105–132.

Leijonmarck, M. and Liljas, A. (1987) Structure of the C-terminal domain of the ribosomal protein L71/L12 from *Escherichia coli* at 1.7 Å. *J. Mol. Biol.*, **195**, 555–580.

Lehninger, A. L. (1981) *Biochemistry: the molecular basis of cell structure and function*, 2nd edn. Worth, New York.

Lehninger, A. L. (1982) *Principles of biochemistry*. Worth, New York.

Leszczynski, J. and Rose, G. (1986) Loops in globular proteins: a novel category of secondary structure. *J. Mol. Biol.*, **234**, 849–855.

Levitt, M. (1976) A simplified representation of protein conformations for rapid simulation of protein folding. *J. Mol. Biol.*, **104**, 59–107.

Levitt, M. and Chothia, C. (1976) Structural patterns in globular proteins. *Nature (London)*, **261**, 552–558.

Lewis, P. N., Go, N., Go, M., Kotelchuck, D. and Scheraga, H. A. (1970) Helix probability profiles of denatured proteins and their correlation with native structures. *Proc. Natl. Acad. Sci. USA*, **65**, 810–815.

Lifson, S. and Sander, C. (1980) Specific recognition in the tertiary structure of β-sheets of proteins. *J. Mol. Biol.*, **139**, 627–639.

Lifson, S. and Warshel, A. (1968) Consistent force field for calculation of vibrational spectra and conformations of some amides and lactam rings. *J. Chem. Phys.*, **49**, 5116–5129.

Liljas, A. and Rossmann, M. G. (1974) X-ray studies of protein interactions. *Annu. Rev. Biochem.*, **43**, 475–507.

Lim, V. I (1974a) Structural principles of the globular organization of protein chains: a stereochemical theory of globular protein secondary structure. *J. Mol. Biol.*, **88**, 857–872.

Lim, V. I. (1974b) Algorithms for prediction of α-helices and β-structural regions in globular proteins. *J. Mol. Biol.*, **88**, 873–894.

Low, B., Preston, H., Sato, A., Rosen, L., Searl, J., Rudko, A. and Richardson, J. S. (1976) Three-dimensional structure of erabutoxin b neurotoxic protein: inhibitior of acetylcholine receptor. *Proc. Natl. Acad. Sci. USA*, **73**, 2991–2994.

Matthews, B. W., Nicholson, H. and Becktel, W. J. (1987) Enhanced protein thermostability from site-directed mutagenesis that decrease the entropy of unfolding. *Proc. Natl. Acad. Sci. USA*, **84**, 6663–6667.

McLachlan, A. D. (1977) Quantum chemistry and protein folding: the art of the possible. *Int. J. Quant. Chem. Mol. Biol.*, **103**, 271–298.

Menendez-Arias, L. and Argos, P. (1989) Engineering protein thermal stability: sequence statistics point to residues substitutions in α-helices. *J. Mol. Biol.*, **206**, 397–406.

Mitchinson, C. and Wells, J.A. (1989) Protein engineering of disulfide bonds in subtilisin BPN'. *Biochemistry*, **28**, 4807–4815.

Momany, F. A., Carruthers, L. M., McGuire, R. F. and Scheraga, H. A. (1974) Intermolecular potentials from crystal data. III. Determination of empirical potentials and application to the packing configurations and lattice energies in crystals of hydrocarbons, carboxylic acids, amines, and amides. *J. Phys. Chem.*, **78**, 1595–1620.

Momany, F. A., McGuire, R. F., Burgess, A. W. and Scheraga, H. A. (1975) Energy parameters in polypeptides. VII. Geometric parameters, partial atomic charges, nonbonded interactions, and intrinsic torsional potentials for the naturally occurring amino acids. *J. Phys. Chem.*, **79**, 2361–2380.

Mozhaev, V. V., Berezin, I. V. and Martinek, K. (1988) Structure–stability relationship in proteins: fundamental tasks and strategy for the development of stabilized enzyme catalysts for biotechnology. *CRC Crit. Rev. Biochem.*, **23**, 235–281.

Nagano, K. (1977) Triplet information in helix prediction applied to the analysis of super-secondary structure. *J. Mol. Biol.*, **109**, 251–274.

Nemethy, G. and Scheraga, H. A. (1977) Protein folding. *Quant. Rev. Biophys.*, **10**, 239–352.

Nozaki, Y. and Tanford, C (1971) The solubility of amino acids and two glycine peptides in aqueous ethanol and dioxane solutions. *J. Biol. Chem.*, **246**, 2211–2217.

Papamokos, E., Weber, E., Bode, W., Huber, R., Empie, M. W., Kato, I. and Laskowski, M., Jr (1982) Crystallographic refinement of Japanese quail ovomucoid, a Kazal-type inhibitor, and model building studies of complexes with serine protease. *J. Mol. Biol.*, **158**, 515–534.

Parry, D. A. D., Grewther, W. G., Fraser, R. D. B. and MacRace, T. P. (1977) Structure of α-keratin: structural implication of the amino acid sequences of the type I and type II chain segments. *J. Mol. Biol.*, **113**, 449–454.

Pauling, L. and Corey, R. B. (1951) Configurations of polypeptide chains with favored orientations around single bonds: two new pleated sheets. *Proc. Natl. Acad. Sci. USA*, **37**, 729–740.

Pauling, L., Corey, R. B. and Branson, H. R. (1951) The structure of proteins: two hydrogen-bonded helical conformations of the polypeptide chain. *Proc. Natl. Acad. Sci. USA*, **37**, 205–211.

Perutz, M. F. (1951) New X-ray evidence on the configuration of polypeptide chains. *Nature (London)*, **167**, 1053–1054.

Pflugrath, J. and Quiocho, F. (1985) Sulfate sequestered in the sulfate-binding protein of *Salmonella typhimurium* is bound solely by hydrogen bonds. *Nature (London)*, **314**, 257–260.

Ponder, J. W. and Richards, F. M. (1987) Tertiary templates for proteins: use of packing criteria in the enumeration of allowed sequences for different structural classes. *J. Mol. Biol.*, **193**, 775–791.

Ponnuswamy, P. K., Prabhakaran, M. and Manavalan, P. (1980) Hydrophobic packing and spatial arrangement of amino acid residues in globular proteins. *Biochim. Biophys. Acta*, **623**, 301–316.

Prevelige, P., Jr and Fasman, G. D. (1989) Chou–Fasman prediction of the secondary structure of proteins: the Chou–Fasman–Prevelige algorithm. In *Prediction of protein structure and the principle of protein conformation* (Fasman, G. D., ed.), pp. 391–416. Academic Press, New York.

Ptitsyn, O. B. and Finkelstein, A. V. (1970a) Connection between the secondary and primary structures of globular proteins. *Biofisika*, **15**, 757–768.

Ptitsyn, O. B. and Finkelstein, A. V. (1970b) Prediction of helical portions of globular proteins according to their primary structure. *Dokl. Akad. Nauk. USSR*, **195**, 221–224.

Ptitsyn, O. B. and Rashin, A. A. (1975) A model of myoglobin self-organization. *Biophys. Chem.*, **3**, 1–20.

Ramachandran, G. N. and Sasisekharan, V. (1968) Conformation of polypeptides and proteins. *Adv. Prot. Chem.*, **23**, 283–437.

Ramachandran, G. N., Ramakrishnan, C. and Sasisekharan, V. (1963) Stereochemistry of polypeptide configurations. *J. Mol. Biol.*, **7**, 95–99.

Ramachandran, G. N., Venkatachalam, C. M. and Krimm, S. (1966) Stereochemical criteria for polypeptide and protein chain conformation. *Biophys. J.*, **6**, 849–872.

Rao, S. T. and Rossmann, M. G. (1973) Comparison of super-secondary structures in proteins. *J. Mol. Biol.*, **76**, 241–256.

Rees, D., Lewis, M. and Lipscomb, W. (1983) Refined crystal structure of carboxypeptidase A at 1.54 Å resolution. *J. Mol. Biol.*, **168**, 367–387.

Rich, A. and Crick, F. (1961) The molecular structure of collagen. *J. Mol. Biol.*, **3**, 483–506.

Richards, F. M. (1974) The interpretation of protein structure: total volume, group volume distributions and packing density. *J. Mol. Biol.*, **82**, 1–14.

Richards, F. M. (1977) Areas, volumes, packing and protein-structure. *Annu. Rev. Biophys. Bioeng.*, **6**, 151–176.

Richardson, J. S. (1976) Handedness of crossover connections in β sheets. *Proc. Natl. Acad. Sci. USA*, **73**, 2619–2623.

Richardson, J. S. (1977) β-sheet topology and the relatedness of proteins. *Nature (London)*, **268**, 495–500.

Richardson, J. S. (1981) The anatomy and taxonomy of protein structure. *Adv. Protein Chem.*, **34**, 167–339.

Richardson, J. S. and Richardson, D. C. (1987) Some design principle: Betabellin. In *Protein engineering*. (Oxender, D. and Fox, C., eds), Ch. 12. Liss, New York.

Richardson, J. S. and Richardson, D. C. (1988) Amino acid preferences for specific locations at the end of α-helices. *Science*, **240**, 1648–1652.

Richardson, J. S. and Richardson, D. C. (1989) Principles and patterns of protein conformation. In *Prediction of protein structure and the principles of protein conformation* (Fasman, G. D., ed.), pp. 1–98. Plenum Press, New York.

Richardson, J. S., Getzoff, E. X. and Richardson, D. C. (1978) The β bulge: a common small unit of nonrepetitive protein structure. *Proc. Natl. Acad. Sci. USA*, **75**, 2574–2578.

Richmond, T. J. and Richards, F. M. (1978) Packing of α-helices: geometrical constraints and contact area. *J. Mol. Biol.*, **119**, 775–791.

Robson, B. and Suzuki, E. (1976) Conformational properties of amino acid residues in globular proteins. *J. Mol. Biol.*, **107**, 327–356.

Robson, B. and Osguthrope, D. J. (1979) Refined models for computer simulations of protein folding: applications to the study of conserved secondary structure and flexible hinge points during the folding of pancreatic trypsin inhibitor. *J. Mol. Biol.*, **132**, 19–51.

Rose, G. (1978) Prediction of chain turns in globular proteins on a hydrophobic basis. *Nature (London)*, **272**, 586–590.

Rose, G., Gaselowitz, A., Lesser, G., Lee, R. and Zehfus, M. (1985) Hydrophobicity of amino acid residues in globular proteins. *Science*, **229**, 834–838.

Salemme, F. (1983) Structural properties of protein β-sheets. *Prog. Biophys. Mol. Biol.*, **42**, 95–133.

Schellman, C. (1980) The α_L conformation at the ends of helices. In *Protein folding* (Jaenicke, R., ed.), pp. 53–61. Elsevier, Amsterdam.

Schiffer, M. and Edmundson, A. (1967) Use of helical wheels to represent the structure of proteins and to identify segments of helical potential. *Biophys. J.*, **7**, 121–135.

Schulz, G. E. and Schirmer, R. H. (1979) *Principles of protein structure*. Springer-Verlag, New York.

Schulz, G. E., Schirmer, R. H., Sachsenheimer, W. and Pai, E. F. (1978) The structure of the flavoenzyme glutathione reductase. *Nature (London)*, **273**, 120–124.

Sheriff, S., Hendrickson, W. A. and Smith, J. L. (1987) Structure of myohemerythrin in the azidomet state at 1.7/1.3 k resolution. *J. Mol. Biol.*, **197**, 273–296.

Shoemaker, K., Kim, P., York, E. and Baldwin, R. (1987) Tests of the helix dipole model for stabilization of α-helices. *Nature (London)* **326**, 563–567.

Sibanda, B. and Thornton, J. (1985) β-Hairpin families in globular proteins. *Nature (London)*, **316**, 170–174.

Smith, W. W., Burnett, R. M., Darling, G. D. and Ludwig, M. L. (1977) Structure of the semiquinone form of flavodoxin from *Clostridium* MP: extension of 1.8 Å resolution and some comparisons with the oxidized state. *J. Biol. Chem.*, **117**, 195–225.

Srinivasan, R. (1976) Helical length distribution from protein crystallographic data. *Ind. J. Biochem. Biophys.*, **13**, 192–193.

Steigeman, W. and Weber, E. (1979) Structure of erythrocruorin in different ligand states refined at 1.4 Å resolution. *J. Mol. Biol.*, **127**, 309–338.

Sternberg, M. J. E. and Thornton, J. M. (1976) On the conformation of proteins: the handedness of the β-strand–α-helix–β-strand unit. *J. Mol. Biol.*, **105**, 367–382.

Sternberg, M. J. E. and Thornton, J. M. (1977) On the conformation of proteins: an analysis of β-pleated sheets. *J. Mol. Biol.*, **110**, 285–296.

Tainer, J., Getziff, E., Beem, K., Richardson, J. and Richardson, D. (1982) Determination and analysis of the 2 Å structure of copper, zinc superoxide dismutase. *J. Mol. Biol.*, **160**, 181–217.

Tainer, J. A., Getziff, E. D., Richardson, J. and Richardson, D. (1982) Determination and analysis of the 2 Å structure of copper, zinc superoxide dismutase. *J. Mol. Biol.*, **160**, 181–217.

Tanford, C. (1963) Contribution of hydrophobic interactions to the stability of the globular conformation of proteins. *J. Am. Chem. Soc.*, **84**, 4240–4247.

Taylor, W. R. and Thornton, J. M. (1984) Recognition of super-secondary structures in proteins. *J. Mol. Biol.*, **173**, 487–514.

Toitskii, G. V. and Zav'yalov, V. P. (1972) Calculation of the conformations of proteins with the aid of a modified nonagram: establishment of the interrelationship between the primary and secondary structures of the polypeptide chain. *J. Mol. Biol.*, **6**, 645–647.

Traub, W. and Piez, K. A. (1971) The chemistry and structure of collagen. *Adv. Prot. Chem.*, **25**, 243–352.

Venkatachalam, C. M. (1968) Stereochemical criteria for polypeptides and proteins. V. Conformation of a system of three linked peptide units. *Biopolymers*, **6**, 1425–1436.

Vonderviszt, F., Matrai, G. and Simon, I. (1986) Characteristic sequential residue environment of amino acids in proteins. *Int. J. Peptide Protein Res.*, **27**, 483–492.

Warshel, A., Levitt, M. and Lifson, S. (1970) Consistent force field for calculation of vibrational spectra and conformations of some amides and lactam rings. *J. Mol. Spectrosc.*, **33**, 84–99.

Watenpaugh, H. D., Sieker, L. C. and Jenson, L. H. (1980) Crystallographic refinement of rubredoxin at 1.2 Å resolution. *J. Mol. Biol.*, **138**, 615–633.

Watson, J. D., Hopkins, N. H., Roberts, J. W., Steitz, J. A. and Weiner, A. M. (1987) *Molecular biology of the gene*, 4th edn. Benjamin, Menlo Park, CA.

Weber, P. and Salemme, F. (1980) Structural and functional diversion in 4-α-helical proteins. *Nature (London)*, **287**, 82–84.

Wetlaufer, D. B. (1973) Nucleation, rapid folding, and globular interchain regions in proteins. *Proc. Natl. Acad. Sci. USA*, **70**, 697–701.

White, J. L., Hackert, M. L., Buehner, M., Adams, M. J., Ford, G. C., Lentz, P. J., Smiley, I. E., Steindel, S. J. and Rossman, M. G. (1976) A comparison of the structures of apo dogfish M_4 lactate dehydrogenase and its ternary complexes. *J. Mol. Biol.*, **102**, 759–779.

Wlodawer, A., Walter, J., Huber, R. and Sjolin, L. (1984) Structure of bovine pancreatic trypsin inhibitor: results of joint neutron and x-ray refinement of crystal form II. *J. Mol. Biol.*, **180**, 301–329.

Wolfenden, R., Andersson, L., Cullis, P. and Southgate, C. (1981) Affinities of amino acid side chains for solvent water. *Biochemistry*, **20**, 849–855.

Wuthrich, K. (1986) *NMR of proteins and nucleic acids*. Wiley, New York.

Wyckof, H. W. (1968) Compensating nature of substitutions in pancreatic ribonucleases. *Brookhaven Symp. Biol.*, **21**, 252–258.

3
Protein stability

The conformational entropy and hydration of the unfolded state of proteins are balanced by specific stabilizing interactions in the folded state. The interactions are those such as the hydrophobic effect, van der Waals forces, electrostatic interactions, H-bonds and disulfide bonds (e.g. Pace, 1975), which were largely identified in the 1950s (Kauzman, 1959). The problem is to determine the contributions of individual amino acids to the stability of a protein; the residues essential for stability, the contributions of physical forces at essential sites, the contributions of critical amino acids depending on environmental conditions, the interactions integrated to produce the observed thermodynamic stability, and the amino acid sequences which destabilize alternative non-functional conformations.

The free energies for the folded and unfolded states are of the order of 10^7 kcal mol^{-1}, including covalent bonds (Baldwin and Eisenberg, 1987). However, the free energy difference between both states is only 5–15 kcal mol^{-1} (Privanov, 1979). The *very accurate* evaluation of such a marginal energy difference, coupled with the inherent structural complexities (see below) therefore appears to be very difficult.

3.1 THERMODYNAMICS OF PROTEIN UNFOLDING

The folded state of proteins is compact and rigid (Chapter 2). However, because of vibrational and conformational entropy, the folded state is indeed flexible (Sturtevant, 1977), suggesting the presence of multiforms of folded protein structure. Alternative folded conformations have actually been detected in a number of proteins (Privalov, 1979; Dlott *et al.*, 1983; Svensson *et al.*, 1986; Smith *et al.*, 1986; Evans *et al.*, 1987).

On unfolding, the polypeptide chain becomes less compact, more highly solvated, and much more flexible (Tanford, 1968). The unfolded chains produced by heat and guanidinium hydrochloride were suggested to be very nearly random coils (Privalov,

1979). A great number of studies, however, have indicated that the unfolded polypeptide produced by various denaturants cannot be an ideal random coil, but with locally ordered conformations, which have been revealed through the structural measurements by optical rotation (Aune et al., 1967), by circular dichroism (Labhart, 1982), by nuclear magnetic resonance (NMR) (Matthews and Westmoreland, 1975; Bierzynski and Baldwin, 1982; Evans et al., 1987), by fluorescence energy transfer (Amir and Haas, 1987; Haas and Amir, 1987), and by kinetic studies of folding (Kim and Baldwin, 1982; Hurle and Matthews, 1987).

When a polypeptide chain in the unfolded state is transferred to aqueous solutions under physiological conditions, the chain begins to fold, because the native state N has a lower free energy than the unfolded state U, due to greatly entropy increase on protein unfolding (section 2.3.1.5). We can presume the following two types of unfolding or folding processes. According to the first classical one, protein folding starts with the polypeptide chain in a completely unfolded random coil state, and proceeds through a set of sequential intermediate forms. According to this scheme, protein unfolding is a non-equilibrium process which can only be investigated kinetically by monitoring structural parameters as a function of time.

Another scheme for protein unfolding is based on the assumption that the polypeptide chain occurs in two states—the N state and the denatured state D—with reversible interconversion. In addition, both N and D are presumed to be single distributions of microstates (Lumry et al., 1966), in spite of the relatively small ensemble of folded conformations and the immense ensemble of unfolded, rapidly interconverting alternatives. According to the first scheme, the protein folding starts from the structureless unfolded state. However, in the second type of protein unfolding (or folding) process a polypeptide chain refolds (or renatures) from the *denatured state* (Tanford, 1968).

The second, reversible denaturation is limited to the simplest monomeric proteins, but it has been shown to be applicable to a number of proteins for understanding the features of protein folding or unfolding (See Chapter 6). For determining protein stability, reversible folding or unfolding experiments have been made, respectively, starting with or ending with the polypeptide chain of local structures produced by denaturation.

3.1.1 Thermodynamic parameters

Reversible denaturation is written as

$$N \rightleftharpoons D$$

From measurement of the equilibrium constant K, which is the ratio of the fraction of molecules in the D state versus the fraction in the N state, the free energy change ΔG on conversion from N to D can be calculated according to equation (2.1) (see section 2.3.1.5). As stated in that section, increasing the number of, or strengthening, the stabilizing interactions in the folded state causes the value of ΔH to become more negative. A decrease in conformational entropy in the unfolded state, i.e. a decrease in the destabilizing forces, will decrease the positive value of $-T\Delta S$. Both the effects increase the negative value of ΔG, i.e. increase the protein stability.

A central feature of the energetics of protein denaturation is that ΔH and ΔS are temperature dependent, as shown in the following equations:

$$\Delta H_T = \Delta H_0 + \Delta C_p(T - T_0)$$
$$\Delta S_T = \Delta S_0 + \Delta C_p \ln(T/T_0)$$

where ΔH_T and ΔS_T are those at temperature T and ΔH_0 and ΔS_0 at a reference temperature T_0. ΔC_p is the difference in heat capacity between the N and D states (Hawkes et al., 1984). ΔC_p ($\Delta H/dT$) gives the heat absorbed by the water of hydrophobic hydration and is about $1-2$ kcal mol^{-1} k^{-1} (Privalov, 1979). In the temperature range 0–80°C, ΔC_p is constant or nearly constant. Because of such large values of ΔC_p, ΔH_T and ΔS_T are strongly dependent on temperature. For example, a temperature change of 1°C causes changes in ΔH_T and in $T\Delta S_T$ of about $1-2$ kcal mol^{-1}. ΔH and ΔS have large compensating values at temperatures at which proteins denature, the mechanism of which, however, is not well understood (Go, 1975; Ueda and Go, 1976).

3.1.2 Determination of ΔG

The value of ΔG is the fundamental measure of the stability of a protein with respect to reversible denaturation. The term protein stability is often used to refer to other reactions, each of which starts with the native state but ends with a different final state—the first type of protein unfolding processes (section 3.1). In the case of irreversible denaturation, for instance, the end state usually involves one or more covalent structural changes (Zales and Klibanov, 1986). In such cases, kinetic, but not thermodynamic, studies can be employed to estimate protein stability. Some examples are shown in Chapters 5 and 6.

A number of different experimental methods can be employed to determine ΔG for comparing the stability of the wild-type and mutant proteins (Hirs and Timashseff, 1986).

Differential scanning calorimetry

The most versatile of these methods is differential scanning calorimetry (DSC), in which the heat absorbed (ΔH) in the transition of N to D and the midpoint temperature (T_m; the midpoint temperature for denaturation) are measured directly (Krishnan and Brandts, 1978). ΔG at temperature T is calculated from the following equation:

$$\Delta G = \Delta H_m - T\Delta S_m + \Delta C_p[T - T_m - T\ln(T/T_m)]$$

where ΔH_m is the enthalpy change and ΔS_m is the entropy change, both at T_m (Hawkes et al., 1984). When this method is applied to mutant proteins, a difference in ΔC_p between wild-type and mutant proteins should be considered (Shortle et al., 1988).

Spectroscopic determination

The loss of ordered structure accompanying denaturation can be monitored spectroscopically by ultraviolet absorption, fluorescence or circular dichroism, which are principally based on the changes of the conformations around aromatic amino acid residues and on the destruction of ordered structures such as α-helices and β-sheets. If the denaturation is approximated by reversible change between the N and D states, the van't Hoff equation

$$\Delta H_{app} = -R\,d(\ln K)/d(1/T)$$

gives the apparent heat absorption ΔH_{app}. In order to calculate ΔG at some reference temperature, ΔC_p is determined by measuring the temperature dependence of ΔH_{app} at different pH values (Privalov, 1979; Hawkes et al., 1984). When the stabilities of the wild-type and its mutant proteins are compared, the wild-type of ΔC_p can be used in the calculation (Shortle and Meeker, 1986). However, this is not always a reliable assumption (Shortle et al., 1988).

Denaturants or pH

The value of ΔG can also be obtained by reversible denaturation with varying concentrations of denaturants or by changing pH. When urea or guanidinium hydrochloride is used as denaturant (Lumry et al., 1966), ΔG has been found to obey the simple linear relationship

$$\Delta G = \Delta G_{water} - m_{den}[\text{denaturant}]$$

where ΔG_{water} is the free energy change in the absence of denaturant. According to Schellman (1978), m_{den} is the difference in solvent-accessible surface area between the N and D states. From the value of m_{den} ($d\Delta G/d[\text{denaturant}]$) determined over the denaturant concentration range where values of ΔG can be measured, ΔG_{water} can be estimated from a linear extrapolation (Pace, 1975). For reversible pH denaturation in a limited pH range, $\Delta G(\text{pH})$ can be determined according to the following equation:

$$\Delta G(\text{pH}) = \Delta G(\text{pH}_0) - 2.303RT(\text{pH} - \text{pH}_0)\Delta v_d$$

Δv_d is the number of hydrogen ions taken up by the protein on acid denaturation or given off on alkaline denaturation, and pH_0 is a reference pH (Hermans and Acampora, 1967).

3.2 STABILIZING FORCES

The approaches for estimating the magnitudes of the interactions or forces which stabilize proteins have traditionally been made using simpler model systems such as organic compounds and oligopeptides (Schulz and Schirmer, 1979), as well as thermophilic proteins (See Chapter 4). However, it has recently been found that genetic and chemical modifications of enzymes are available for such approaches (See Chapters 5 and 6). Genetic modification especially, i.e. site-directed mutagenesis (Chapter 6), is indeed useful for introducing specific changes in the amino acid sequence of proteins for understanding the quantitative as well as qualitative roles of interactions in protein stability and stabilization. Data on several classes of possible

3.2 Stabilizing Forces

interactions for stabilizing proteins are summarized in this section. More detailed data will be presented in Chapters 5 and 6.

3.2.1 Dispersion forces

During protein folding, water is transferred from the surface of polypeptide chains to the relatively open bulk phase, and the atoms that form the protein interior through dispersion or van der Waals forces become as densely packed as small organic molecules in the crystal state (Klapper, 1971; Chothia, 1975; Richards, 1977). The packing of proteins is not strictly as compact as that of small organic solids, because of the greater flexibility or the irregularities in the packing of individual side chains with different neighbors of proteins (Hvidt, 1975; Narayana and Argos, 1984).

Amino acid substitutions can alter protein stability by changing interior packing, although quantitative correlations have not yet been established. The lack of stabilization by an engineered disulfide bond in dihydrofolate reductase is due to a reduction in tertiary contact surface (Villafranca et al., 1987). Comparison of structure and stability of the wild-type and engineered proteins of T4 lysozyme showed the contribution of van der Waals and hydrophobic contacts of the γ-methyl group of Thr 157 in the wild-type protein to 0.7 kcal mol^{-1} stabilization at pH 2 and 42°C (Alber et al., 1987b). On replacing Ala 146 with Thr of the enzyme the cavity at this site is eliminated, and the same cavity is expanded by the replacement of Met 102 by Thr (Grutter et al., 1983). Both amino acid substitutions destabilize the protein. Creating a cavity in bacterial ribonuclease by site-directed mutagenesis also reduces the stability (Kellis et al., 1988). These results clearly demonstrate the relation of the cavity in the protein interior to stability.

3.2.2 Conformational entropy

Covalent cross-links such as disulfide bonds are considered to stabilize proteins by reducing the conformational entropy of the unfolding chain, which depends on the length of the loop formed by a single cross-link. Flory (1956) formulated the chain entropy (ΔS) of disulfide cross-links as a function of the length of the loop as follows:

$$\Delta S = 0.75 \sigma R \ln(n' + 3)$$

where σ is the number of connected chains (twice the number of cross-links) and n' is the number of amino acid residues of each loop (Nakajima and Scheraga, 1961). Recently, Matsumura et al. (1989) reported the correlation between the loop size of engineered disulfide bonds and stability of the mutant proteins in T4 lysozyme. The cross-link may also have an effect on the folded state.

Chemical cross-linkage stabilizes ribonuclease A by about 5 kcal mol^{-1} at 53°C (Lin et al., 1984). This stabilizing effect is entropic; i.e. consistent with reduction of the conformational entropy of the unfolded state. Kinetic analysis, however, showed that the cross-link specifically reduces the rate of unfolding (Lin et al., 1985). Matthews and Hurle (1987) then considered that the cross-link may differentially affect the folded state.

The N- and C-termini of bovine pancreatic trypsin inhibitor (BPTI), which are linked by an ion pair in the wild-type protein, was chemically cross-linked by

Goldenberg and Creighton (1984). The modified and wild-type proteins exhibited similar stability. This suggests that the stabilizing effect by entropy decrease of the unfolded state was compensated by the destabilization of the folded state.

The idea of *effective concentration* (the ratio of the intramolecular and intermolecular association constants for two groups) has recently been proposed (Creighton, 1983a, 1983b; Creighton and Goldenberg, 1984; Goldenberg, 1985). If the groups have higher effective concentrations, i.e. their interactions are more favored, in the folded state than in the unfolded state, they contribute to protein stability. The three natural disulfides in BPTI have similar stabilities relative to an added disulfide exchange reagent in the unfolded state, and these effective concentrations are about 10^{-2} M. In the folded state the concentrations range from 2.3×10^2 M to 4.6×10^5 M (Creighton, 1983a). The effective concentration of each disulfide bond in the folded state was shown to be correlated qualitatively with the change in the melting temperature of the protein caused by reduction of the disulfide.

It may then be suggested that the stabilizing contribution of a disulfide bond to proteins is determined by the loop size formed and also by the compatibility of the cross-link with the folded structure. For a given loop size, the most effective cross-link may be formed between groups that are rigidly held in an optimum orientation by the folded state (Creighton, 1983a; Goldenberg, 1985).

To estimate the contribution of the disulfide bond to protein stability, engineering of disulfide bond(s) was attempted in proteins without any disturbance of the conformational restrictions observed in naturally occurring proteins (Richardson, 1981). Very few pairs of Cys residues—either or both engineered—however, can be joined by a disulfide bond with standard geometry (Pabo and Suchaneck, 1986). Introduction of disulfide bond(s) in subtilisin (Perry and Wetzel, 1984; Wells and Powers, 1986: Pantoliano *et al.*, 1987; Wetzel *et al.*, 1988; Mitchinson and Wells, 1989) and in dihydrofolate reductase (Villafranca *et al.*, 1987) produced complex effects (see Chapter 6). In most cases, the Cys introduction for an engineered disulfide bond was destabilizing. For example, the lysozyme containing an engineered disulfide between positions 3 and 97 is $1.8 \, \text{kcal mol}^{-1}$ more stable than the reduced mutant, but only $1.2 \, \text{kcal mol}^{-1}$ more stable than the wild-type protein (Wetzel *et al.*, 1988).

Residues that do not introduce disulfides may also affect the conformational entropy of the unfolded state. Since Gly has no side chain, its conformational freedom is higher than those of other amino acid residues. From the degree of freedom of a polypeptide backbone, it was suggested that removal of a Gly residue could destabilize the unfolded state by $0.8 \, \text{kcal mol}^{-1}$ at 65°C (Nemethy *et al.*, 1966). Amino acid substitutions which introduce β-branched or bulkier side chains may also restrict main chain rotations in the unfolded state (Nemethy *et al.*, 1966). The increases in stability, although below 1 kcal/mol, were caused by replacing Gly with other amino acids or by substitution of Pro compatible with the folded state of T4 lysozyme (Matthews *et al.*, 1987), λ repressor (Hecht *et al.*, 1984a, 1984b, 1985) and thermophilic neutral protease (Imanaka *et al.*, 1986). The increases in stability of λ repressor and neutral protease were contributed by enhanced helix propensity in helical regions (section 2.7.1) (Chou and Fasman, 1978). However, entropic effects may contribute to the observed tendency of substituting Gly to stabilize model helices. In addition to the entropic effects on the unfolded state, the substitutions could increase stability of proteins by increasing their rigidity, and consequently the effective concentration of neighboring groups in the folded structure (Alber et al., 1987a, 1987b).

3.2.3 Hydrophobic interactions

One of the most important interactions for stabilizing proteins is hydrophobic effects (section 3.1).

The free energy of transferring side chains from water to ethanol (Tanford, 1962; Nozaki and Tanford, 1971) is correlated with their average extent of burial in protein structure, so that the solvent transfer model of Tanford can be applied to quantitatively estimate protein stability (Rose et al., 1985; Miller et al., 1987; Lawrence et al., 1987). The free energy of stabilization by hydrophobic forces has recently been found by summing the different contributions of individual atoms (Eisenberg and McLachlan, 1986; Abraham and Leo, 1987).

For hen egg lysozyme (Baldwin, 1986), the temperature-dependent part of ΔH and ΔS is largely due to the hydrophobic effect. The temperature-independent fraction of ΔH and ΔS is raised from the contribution by other interactions. The temperature-independent part of ΔH is large and favors folding, because of non-covalent interactions in the folded state. The temperature-independent part of ΔS is also large, but favors unfolding, because of the conformational entropy of the unfolded state. The free energy of stabilization by the hydrophobic forces increased with temperature from 10 to 100°C (Kauzman, 1959), and it reached a maximum at about 112°C, if ΔC_p is constant. At higher temperatures the destabilizing contribution by conformational entropy of the unfolded state ($T\Delta S$) increases more rapidly with temperature than the stabilizing contribution by the hydrophobic effect, so that the protein unfolds or denatures. It should be noted that the analysis was limited by the assumption made to determine ΔC_p and by the possibility that the hydrophobic scales used may not be directly applicable to protein folding.

As a model of stabilizing effects of hydrophobic side chains, the transfer energy of hydrophobic groups from water to hydrophobic environments (Tanford, 1962; Nozaki and Tanford, 1971) has been widely employed. However, the relevance of such transfer experiments has still been questioned on theoretical and experimental grounds (Brandts et al., 1970; Klapper, 1971; Zipp and Kauzmann, 1973; Hvidt, 1975; Chothia, 1976; Richards, 1977; Tanford, 1980; Wolfenden et al., 1981; Creighton, 1983b; Lee, 1985). For instance, because of differences in the energy of cavity formation in the protein interior (Lee, 1985), hydrophobicity may vary from site to site and from protein to protein. The relationship between hydrocarbon transfer free energy in model system and hydrophobic stabilization of protein has not yet been fully unerstood.

The qualitative application of transfer experiments to protein stability was shown with the engineered tryptophan synthase α subunit (Yutani et al., 1984, 1987) and kanamycin nucleotidyltransferase (Matsumura et al., 1988a). However, the stability of T4 lysozyme was directly related to the hydrophobicity of the substituted side chain (Matsumura et al., 1988b). In addition, the hydrophobic stabilization was proportional to the reduction of surface area accessible to solvent on folding.

Removal of hydrophobic interactions by appropriate amino acid substitutions of ribonuclease from *Bacillus amyloliquefaciens* introduced cavities into the enzyme (Kellis *et al.*, 1988). Creation of a cavity with a size of a —CH_2— group destabilized the enzyme by 1.1 kcal mol^{-1}.

Hydrophobicity scales of amino acid residues are shown in Table 4.1.

3.2.4 Hydrogen bonding

Temperature-independent ΔH is contributed by the interactions other than the hydrophobic effect, as described previously, and it is thought to come largely from H-bonds. Highly refined X-ray crystal analysis of proteins shows that H-bonding geometry is indeed variable, and almost all H-bond donors and acceptors are paired (Baker and Hubbard, 1984).

Some model compounds which form intermolecular H-bonds in aqueous solution have been used to estimate the contribution of H-bonding to protein stability. However, because of competition with water the compounds associate only weakly in water, and different model systems have provided different estimates of the enthalpy of H-bond formation. For instance, studies of the dimerization of urea or δ-valerolactum in water provided a value of -1.5 kcal mol^{-1} or more for the heat of H-bond formation (Schellman, 1955a, 1955b; Susi *et al.*, 1964), but with the model peptide methylacetamide ΔH for H-bond formation was nearly zero (Klotz and Franzen, 1962).

On folding, water molecules which are H-bonded with unfolded polypeptide chain are released to interact with each other, and specific intermolecular H-bounds are formed in the protein interior, and specific sites which are H-bonded with water are formed on the folded state. If the number of H-bonds in the folded and unfolded states is equal and if the effective concentration of H-bonding groups is equal in both states, no energy difference may be observed between the folded and unfolded states, as actually reported by Klotz and Franzen (1962). The energies between H-bonds in the folded and unfolded states, however, will be different, because of the differences in the average geometry, the entropy of formation, and the average number of interacting partners of H-bonds (Page and Jencks, 1971; Creighton, 1983a, 1983b; Goldenberg, 1985; Stahl and Jencks, 1986). In addition, all possible H-bonds of proteins are not formed in the aqueous environment. The data obtained from classical model experiments, therefore, cannot be simply applied to estimate the contribution of H-bonding.

Studies on the H-bonding energy in biological materials were recently made with protein-ligand complexes and nucleic acids (Page, 1984; Fersht *et al.*, 1985; Fersht, 1987). H-bonds between uncharged donors and acceptors were calculated to contribute 0.5–1.8 kcal mol^{-1}, and those between charged groups 6 kcal mol^{-1}.

Recent developments of protein engineering (Chapter 6) have made it possible to estimate quantitatively as well as qualitatively the contribution of H-bonding to protein stability by site-directed mutagenesis. The Thr at position 157 in T4 lysozyme was replaced with various amino acids, and their X-ray crystal structures and thermodynamic stabilities were compared (Grutter *et al.*, 1987; Alber *et al.*, 1987b). The studies confirmed that unpaired H-bond donors or acceptors sequestered from solvent in the folded state are destabilizing. Any conclusive evidence for excess free energy for protein stabilization by specific H-bonds, however, was difficult to deduce, because the amino acid substitutions simultaneously altered several interactions in the folded state (see also Chapter 6).

The surface-exposed Pro at position 86 in T4 lysozyme was replaced with ten other amino acids (Alber et al., 1988). The side chains of Ser and Cys substituted at this site formed a unique tertiary H-bond to the side chain of Gln 122. However, these mutant proteins exhibited the same stability as that of the wild-type protein, indicating no correlation of the added H-bond with stability. Residues at position 86 have higher than average mobility and surface accessibility, so that the intramolecular H-bond may be equivalent to the local interactions with water. The high mobility of the side chains at this site may reflect a significant loss of entropy when they form the intramolecular bond.

3.2.5 Electrostatic interactions

Protein folding alters the effective dielectric constant around each ion and produces the two different regions: the protein surface exposed to high dielectric constant and the protein interior of lower dielectric constant. Thus, the unique changes in the solvation of specific groups and the differences in the entropy of formation of specific interactions are induced on folding.

Charged groups, including the peptide dipoles, are generally surrounded by charges of the opposite sign (Wada and Nakamura, 1981; Thornton, 1982; Barlow and Thornton, 1983). Only a third of the charged residues exist as ion pairs, and about 80% of the ion pairs are between residues in different parts of secondary structure (Barlow and Thornton, 1983). Unexpectedly, 17% of the ion pairs are buried, and they play identifiable functional roles. Indeed 20% of the buried charged groups do not form ion pairs but are solvated (Rashin and Honig, 1984). This indicates the importance of the self-energy of ionic groups (Russell and Warshel, 1985; Gilson et al., 1985; Warshel, 1987).

Many results indicating the relation of ionic interactions with protein stability as well as with protein folding have been reported. The contribution of additional salt bridges was first suggested to enhance thermostability of thermophilic proteins (Perutz and Raidt, 1975; Perutz, 1978). Charged residues, even at the sites of mobility and water accessibility, can be sensitive to destabilizing mutation (Hecht et al., 1983; 1984a, 1984b; Pakula et al., 1986, Alber et al., 1987a). Charged groups also affected the stability of isolated helices (Ihara et al., 1982; Shoemaker et al., 1985; Mitchinson and Baldwin, 1986; Marqusee and Baldwin, 1987). In tryptophan synthase α-subunit the replacement of Arg 211 by Glu increased the folding rate by a factor of about 70 (Matthews et al., 1983; Beasty et al., 1987).

In addition, the stabilizing contributions of ion pairs to protein stability have been quantitatively estimated. In α- and δ-chymotrypsins the stabilizing energy for pH inactivation of the buried ion pair between Arg 194 and the α-amino group of Ile 16 was estimated to be 2.9 kcal mol^{-1} (Fersht, 1971). Recently, Fersht (1987) and his co-workers (1985) showed the contribution of 3–6 kcal mol^{-1} to ligand-binding energies. A solvent-exposed salt bridge between the termini of BPTI can stabilize the protein about 1 kcal mol^{-1} in 6 M guanidinium hydrochloride at pH 6 and 75°C (Brown et al., 1978). In T4 lysozyme a new ion pair formed with the substituted Arg at position 157 (Alber et al., 1987b) contributed about 1 kcal mol^{-1} to the thermal stability of the variant (Alber et al., unpublished results).

The unpaired charged groups at specific sites can affect protein stability (Wagner et al., 1979; Matthews et al., 1980; Ahern et al., 1987; Alber et al., 1988). In the α-subunit of tryptophan synthase, replacement of Gly 211 by either Glu or Arg increases stability (Matthews et al., 1980). Charged residues at helix termini were also shown to correlate to the stability of ribonuclease S (Mitchinson and Baldwin, 1986) and of T4 lysozyme (Sturtevant, 1987). The guanidinium groups of Arg 96 in T4 lysozyme forms H-bonds with two carbonyl groups at the C-terminus of a helix (Grutter et al., 1979). Replacement of Arg 96 by His reduced the free energy of stabilization by 3.5 kcal mol^{-1} (Sturtevant, 1987).

Ionic interactions are affected by the changes in environmental pH (Becktel and Schellman, 1987) and in local effective dielectric constants (Cronin et al., 1987). The stability of a protein depends on the number of protons bound or released on denaturation (Tanford, 1968, 1970; Privalov, 1979; Becktel and Schellman, 1987). Protein stability changes smoothly with pH, suggesting the compensating interactions around ionizable groups. Thus, in sperm whale myoglobin stabilizing and destabilizing interactions were redistributed over the charge pairs as the pH was changed (Garcia-Moreno et al., 1985). Since the polar groups are distributed within a protein the local dielectric can be very high (Macgregor and Weber, 1986), and local differences in polarity are also proposed (Warshel, 1987). It was then considered that local difference in polarity or in dielectric contribute to the surprising 2.7 kcal mol^{-1} difference in the stability on inverting the Asp–Arg ion pair to an Arg–Asp pair in aspartate amino tansferase (Cronin et al., 1987).

Even non-polar side chains may participate in weakly polar interactions (Burley and Petsklo, 1985). Aromatic groups in proteins tend to pack edge to face. Such arrangement of two aromatic rings may induce short-range dispersion forces and a long-range electrostatic attraction between the electron-poor aromatic hydrogen and the electron-rich π orbitals (Burley and Petsko, 1985, 1988).

3.2.6 Structural features

Proteins can be very tolerant of amino acid substitutions, as exemplified in the *Escherichia coli lac* repressor mutants (Miller et al., 1979; Miller and Schmeissner, 1979; Miller, 1984) and in globin families (Perutz et al., 1965; Perutz and Lehmann, 1968; Go and Miyazawa, 1980). This may be explained by the following reasons: only a fraction of the residues in a protein contribute to stability (Alber et al., 1987a) and structural adjustments mitigate the intrinsic effects of amino acid substitutions, even at critical sites (Alber et al., 1987b, 1988). A large number of randomly induced mutants that alter protein stability, chiefly destabilizing, have been collected with hemoglobins (Fermi and Perutz, 1981), T4 lysozyme (Streisinger et al., 1961; Alber et al., 1987a) and the *cI* and *cro* repressors of phase λ (Hecht et al., 1983, 1984a; Pakula et al., 1986). A number of general conclusions have been drawn from the structural studies on these mutant proteins, although the characteristics of stability differ from system to system.

The structural and thermodynamic studies on the temperature-sensitive mutant proteins revealed that any specific amino acid substitutions, with respect to charge, size, polarity, hydrophobicity or hydrogen bonding capacity, are not necessary for protein stability. All the non-covalent interactions or forces as described above may contribute to protein stability.

Genetic studies, however, suggest that only some of the amino acids greatly contribute to stability, and that many interactions are not necessarily stabilizing. Most critical amino acids which destabilize λcI repressor (Hecht et al., 1983, 1984a) and T4 lysozyme (Alber et al., 1987a) were relatively rigid and inaccessible to water in the folded state. Substitutions at mobile and exposed sites generally have little effect on stability. Similar results were obtained with the λ cro repressors, with some exceptions (Pakula et al., 1986). Side chains with a large amount of buried non-polar surface area are sensitive to destabilizing substitutions, indicating the importance of van der Waals and hydrophobic contacts. Even small buried residues, however, form part of the target for temperature-sensitive mutations. This cannot be reasonably explained by a linear relationship between surface area and hydrophobic stabilizations, because small side chains have small surface area. This may support the view that groups that become surrounded by atoms during protein folding can form specific stabilizing interactions.

The rigid and buried residues are involved in the part of the folded state which is least able to relax structurally to compensate for amino acid substitutions. Significant changes on the folded state may then produce large changes in its thermodynamic stability. The conformational diversity of the unfolded state, on the other hand, may allow it to compensate more completely for changes in amino acid

Table 3.1—Factors to consider in stabilizing proteins.

Interaction
   Disulfide bond
   Hydrogen bond
   Electrostatic interaction
   Hydrophobic interaction

Conformational factors
   Stability of α-helix
   Compact packing
   Internal hydrophobicity
   Entropic stability

Protection
   Deamination of carboxylamide
   Oxidation of sulfhydryl groups
   Intramolecular S—H/S—S exchange
   Oxidation of tryptophan or methionine

These factors were deduced by comparing the structures of proteins of different stabilities: those isolated from mesophiles and thermophiles, wild-type and amino acid-substituted proteins, and wild-type and chemically modified proteins. The data described in each chapter are summarized.

sequence, and the free energy of the state may be less sensitive to substitutions. Analysis of mutant proteins would be simplified if the effects on the folded state are dominant. If mutations cause small changes in stability, the effects on the unfolded state would become more important. The controversial results, however, were reported on the effects of substitutions on the unfolded state (Shortle and Meeker, 1986; Beasty et al., 1986; Matthews and Hurle, 1987).

Protein folding is cooperative, but multiple amino acid substitutions generally have additive effects on stability (Shortle and Lin, 1985; Matsumura et al., 1986; Becktel and Schellman, 1987). Non-additive effects were observed if one of the substitutions altered ΔC_p (Wetzel et al., 1988). Substitutions should synergistically alter the contribution of neighboring interactions and lead to non-additive effects even in the absence of structural shifts (Creighton, 1983a, 1983b; Goldenberg and Creighton, 1985). The additivity of substitutions implies that cooperative interactions are highly localized. This supports the simplifying assumption that protein stability may be a sum of local stabilization (Menendez and Argos, 1989). Multiple substitutions might be engineered to increase stability up to any desired level.

Factors to consider in stabilizing proteins are summarized in Table 3.1.

REFERENCES

Abraham, D. J. and Leo, A. J. (1987) Extension of the fragment method to calculate amino acid zwitterion and side chain partition coefficient. *Proteins Struct. Funct. Genet.*, **2**, 130–152.

Ahern, T. J., Casal, J. I., Petsko, J. A. and Klibanov, A. M. (1987) Control of oligomeric enzyme thermostability by protein engineering. *Proc. Natl. Acad. Sci. USA*, **84**, 675–679.

Alber, T., Sun, D.-P., Nye, J. A., Muchmore, D. C. and Matthews, B. W. (1987a) Temperature-sensitive mutations of bacteriophage T4 lysozyme occur at sites of low mobility and low solvent accessibility in the folded protein. *Biochemistry*, **26**, 3754–3758.

Alber, T., Sun, D.-P., Wilson, K., Wozniak, J. A., Cook, S. P. and Matthews, B. W. (1987b) Contributions of hydrogen bonds of Thr 147 to the thermodynamic stability of phage T4 lysozyme. *Nature (London)*, **330**, 41–46.

Alber, T., Bell, J. A., Sun, D.-P., Nicholson, H., Wozniak, J. A., Cook, S. P. and Matthews, B. W. (1988) Replacements of Pro 86 in phage T4 lysozyme extend an α-helix but do not alter protein stabilty. *Science*, **239**, 631–635.

Amir, D. and Haas, E. (1987) Estimation of intramolecular distances in bovine pancreatic trypsin inhibitor by site-specific labeling and nonradiative excitation energy-transfer measurements. *Biochemistry*, **26**, 2162–2175.

Aune, K. C., Salahuddin, A., Zarlengo, M. H. and Tanford, C. (1967) Evidence for residual structure in acid- and heat-denatured proteins. *J. Biol. Chem.*, **242**, 4486–4489.

References

Baker, E. N. and Hubbard, R. E. (1984) Hydrogen bonding in globular proteins. *Prog. Biophys. Mol. Biol.*, **44**, 97–179.

Baldwin, R. L. (1986) Temperature dependence of the hydrophobic interaction in protein folding. *Proc. Natl. Acad. Sci. USA*, **83**, 8069–8072.

Baldwin, R. L. and Eisenberg, D. (1987) Protein stability. In *Protein engineering* (Oxender, D. L. and Fox, C. F., eds), pp. 127–148. Liss, New York.

Barlow, D. J. and Thornton, J. M. (1983) Ion-pairs in proteins. *J. Mol. Biol.*, **168**, 867–885.

Beasty, A. M., Hurle, M. R., Manz, J. T., Stackhouse, T., Onuffer, J. J. and Matthews, C. R. (1986) Effects of the Phe 22 to Leu, Glu 49 to Met, Gly 234 to Asp, and Gly 234 to Lys mutations on the folding and stability of the α-subunit of tryptophan synthase from *Escherichia coli*. *Biochemistry*, **25**, 2965–2974.

Beasty, A. M., Hurle, M., Manz, J. T., Stackhouse, T. and Matthews, C. R. (1987) Mutagenesis as a probe of protein folding and stability. In *Protein engineering* (Oxender, D. L. and Fox, C. F., eds), pp. 91–102. Liss, New York.

Becktel, W. J. and Schellman, J. A. (1987) Protein stability curves. *Biopolymers*, **26**, 1859–1877.

Bierzynski, A. and Baldwin, R. L. (1982) Local secondary structure in ribonuclease A denatured by guanidinium HCl near 1°C. *J. Mol. Biol.*, **162**, 172–186.

Brandts, J. E., Oliveira, R. J. and Westort, C. (1970) Thermodynamics of protein denaturation: effect of pressure on the denaturation of ribonuclease A. *Biochemistry*, **9**, 1038–1048.

Brown, L. R., DeMarco, A., Richard, R., Wagner, G. and Wutrich, K. (1978) The influence of a single salt bridge on static and dynamic features of the globular solution conformation of the bovine pancreatic trypsin inhibitor. *Eur. J. Biochem.*, **88**, 87–95.

Burley, S. K. and Petsko, G. A. (1985) Aromatic–aromatic interaction: a mechanism of protein structure stabilization. *Science*, **229**, 23–28.

Burley, S. K. and Petsko, G. A. (1988) Weakly polar interactions in proteins. *Adv. Protein Chem.*, **39**, 125–189.

Chothia, C. (1975) Structural invariants in protein folding. *Nature (London)*, **254**, 304–308.

Chothia, C. (1976) The nature of the accessible and buried surfaces in proteins. *J. Mol. Biol.*, **105**, 1–14.

Chou, P. Y. and Fasman, G. D. (1978) Prediction of the secondary structure of proteins from their amino acid sequence. *Adv. Enzymol.*, **47**, 45–148.

Creighton, T. E. (1983a) An empirical approach to protein conformation, stability and flexibility. *Biopolymers*, **22**, 49–58.

Creighton, T. E. (1983b) *Proteins*. Freeman, New York.

Creighton, T. E. and Goldenberg, D. P. (1984) Kinetic role of a metastable native-like two-disulfide species in the folding transition of bovine pancreatic trypsin inhibitor. *J. Mol. Biol.*, **179**, 497–526.

Cronin, C. N., Malcolm, B. A. and Kirsh, J. F. (1987) Reversal of substrate specificity by site-directed mutagenesis of aspartate amino transferase. *J. Am. Chem. Soc.*, **109**, 2222–2223.

Dlott, D. D., Frauenfelder, H., Langer, P., Roader, H. and Dilorio, E. E. (1983) Nanosecond flash photolysis study of carbon monoxide binding to the β chain of hemoglobin Zurich [β63(E7)His to Arg]. *Proc. Natl. Acad. Sci. USA*, **80**, 6293–6243.

Eisenberg, D. and McLachlan, A. D. (1986) Solvation energy in protein folding and unfolding. *Nature (London)*, **319**, 199–203.

Evans, P. A., Dobson, C. M., Kautz, R. A., Hatfull, G. and Fox, R. O. (1987) Proline isomerism in staphylococcal nuclease characterized by NMR and site-directed mutagenesis. *Nature (London)*, **239**, 266–268.

Fermi, G. and Perutz, M. F. (1981) *Haemoglobin and myoglobin*. Clarenden Press, Oxford.

Fersht, A. R. (1971) Conformational equilibria and the salt bridge in chymotrypsin. *Cold Spring Harbor Symp. Quant. Biol.*, **36**, 71–73.

Fersht, A. R. (1987) The hydrogen bond in molecular recognition. *Trends Biochem. Sci.*, **12**, 301–304.

Fersht, A. R., Shi, J.-P., Knill-Jones, J., Lowe, D. M., Wilkinson, A.J., Blow, D. M., Brick, P., Carter, P., Waye, M. M. Y. and Winter, G. (1985) Hydrogen bonding and biological specificity by protein engineering. *Nature (London)*, **314**, 235–238.

Flory, J. G. (1956) Theory of elastic mechanisms in fibrous proteins. *J. Am. Chem. Soc.*, **78**, 5222–5235.

Garcia-Moreno, E. G., Chen, L. N., March, K. L., Gurd, R. S. and Gurd, F. R. N. (1985) Electrostatic interactions in sperm whale myoglobin. *J. Biol. Chem.*, **260**, 14070–14082.

Gilson, M. K., Rashin, A., Fine, R. and Honig, B. (1985) On the calculation of electrostatic interactions in proteins. *J. Mol. Biol.*, **183**, 503–516.

Go, N. (1975) Theory of reversible denaturation of globular proteins. *Int. J. Peptides*, **7**, 313–323.

Go, N. and Miyazawa, S. (1980) Relationship between mutability, polarity and exteriority of amino acid residues in protein evolution. *Int. J. Peptides*, **15**, 211–224.

Goldenberg, D. P. (1985) Dissecting the roles of individual interactions in protein stability: lessons from a circularized protein. *J. Cell Biochem.*, **29**, 321–335.

Goldenberg, D. P. and Creighton, T. E. (1984) Folding pathway of a circular form of bovine pancreatic trypsin inhibitor. *J. Mol. Biol.*, **179**, 527–545.

Goldenberg, D. P. and Creighton, T. E. (1985) Energetics of protein structure and folding. *Biopolymers*, **24**, 167–182.

Grutter, M. G., Hawkes, R. B. and Matthews, B. W. (1979) Molecular basis of thermostability in the lysozyme of bacteriophage T4. *Nature (London)*, **277**, 667–668.

Grutter, M. G., Weaver, L. H., Gray, T. M. and Matthews, B. W. (1983) Structure, function, and evolution of the lysozyme from bacteriophage T4 lysozyme. In *Bacteriophage T4* (Matthews, C. K., Kutter, E. M., Mosig, G. and Berget, P. M., eds), pp. 356–360. American Society for Microbiology, Washington.

Grutter, M. G., Gray, T. M., Weaver, L. H., Alber, T., Wilson, K. and Matthews, B. W. (1987) Structural studies of mutants of the lysozyme of bacteriophage T4: the temperature-sensitive mutant protein Thr 157 to Ile. *J. Mol. Biol.*, **197**, 315–329.

Haas, E. and Amir, D. (1987) BPTI has a compact structure when the disulfide bonds are reduced. *J. Cell. Biochem.*, **11C**, 214.

Hawkes, R., Grutter, M. G. and Schellman, J. (1984) Thermodynamic stability and point mutations of bacteriophage T4 lysozyme. *J. Mol. Biol.*, **175**, 195–212.

Hecht, M. H., Nelson, H. C. M. and Sauer, R. T. (1983) Mutation in λ repressor's amino-terminal domain: implications for protein stability and DNA binding. *Proc. Natl. Acad. Sci. USA*, **80**, 2676–2680.

Hecht, M. H., Sturtevant, J. M. and Sauer, R. T. (1984a) Effect of amino acid replacements on the thermal stability of the NH_2-terminal domain of phage λ repressor. *Proc. Natl. Acad. Sci. USA*, **81**, 5658–5689.

Hecht, M. H., Sturtevant, J. M. and Sauer, R. T. (1984b) Stabilization of λ repressor against thermal denaturation by site-directed Gly to Ala changes in α-helix. *Proc. Natl. Acad. Sci. USA*, **81**, 5685–5689.

Hecht, M. H., Hehir, K. M., Nelson, H. C. M., Sturtevant, J. M. and Sauer, R. T. (1985) Increasing and decreasing protein stability: effects of revertant substitutions on the thermal denaturation of phage λ repressor. *J. Cell. Biochem.*, **29**, 217–224.

Hermans, J., Jr and Acampora, G. (1967) Reversible denaturation of sperm whale myoglobin. II. Thermodynamic analysis. *J. Am. Chem. Soc.*, **89**, 1547–1552.

Hirs, C. H. and Timasheff, S. N. (eds) (1986) Section I. Unfolding and refolding of proteins. *Methods Enzymol.*, **131**, 1–280.

Hurle, M. R. and Matthews, C. R. (1987) Proline isomerization and the slow folding reactions of the α subunit of tryptophan synthease from *Escherichia coli*. *Biochim. Biophys. Acta*, **913**, 179–184.

Hvidt, A. (1975) A discussion of pressure–volume effects in aqueous protein solution. *J. Theor. Biol.*, **50**, 245–252.

Ihara, S., Ooi, T. and Takahashi, S. (1982) Effects of salts on the nonequivalent stability of the α-helices of isomeric block copolypeptides. *Biopolymers*, **21**, 131–145.

Imanaka, T., Shibazaki, M. and Aiba, M. (1986) A new way of enhancing the thermostability of proteins. *Nature (London)*, **324**, 695–697.

Kauzman, W. (1959) Some factors in the interpretation of protein denaturation. *Adv. Protein Chem.*, **14**, 1–63.

Kellis, J. T., Jr, Nyberg, K., Sali, D. and Fersht, A. R. (1988) Contribution of hydrophobic interactions to protein stability. *Nature (London)*, **333**, 784–786.

Kim, P. S. and Baldwin, R. L. (1982) Specific intermediates in the folding reactions of small proteins and the mechanism of protein folding. *Annu. Rev. Biochem.*, **51**, 459–489.

Klapper, M. H. (1971) The nature of the protein interior. *Biochim. Biophys. Acta*, **229**, 557–566.

Klotz, I. M. and Franzen, J. S. (1962) Hydrogen bonds between model peptide groups in solution. *J. Am. Chem. Soc.*, **84**, 3461–3466.

Krishnan, K. S. and Brandts, J. F. (1978) Scanning calorimetry. *Methods Enzymol.*, **49**, 3–14.

Labhart, A. M. (1982) Secondary structure in ribonuclease I. Equilibrium folding transitions seen by amide circular dichroism. *J. Mol. Biol.*, **157**, 331–355.

Lawrence, C., Auger, I. and Mannella, C. (1987) Distribution of accessible surface areas of amino acids in globular proteins. *Proteins Struct. Funct. Genet.*, **2**, 153–167.

Lee, B. (1985) The physical origin of the low solubility of nonpolar solutes in water. *Biochemistry*, **24**, 813–823.

Lin, S.H., Konishi, Y., Denton, M. E. and Scheraga, H. A. (1984) Influence of an extrinsic crosslink on the folding pathway of ribonuclease A: conformational and thermodynamic analysis of cross-linked (lysine7,lysine41)-ribonuclease A. *Biochemistry*, **23**, 5504–5512.

Lin, S. H., Konishi, Y., Nall, B. T. and Scheraga, H. A. (1985) Influence of an extrinsic cross-link on the folding pathway of ribonuclease A: kinetics of folding–unfolding. *Biochemistry*, **24**, 2680–2686.

Lumry, R., Biltonen, R. and Brandts, J. F. (1966) Validity of the 'two-state' hypothesis for conformational transitions of proteins. *Biopolymers*, **4**, 917–944.

Macgregor, R. B. and Weber, G. (1986) Estimation of the polarity of the protein interior by optical spectroscopy. *Nature (London)*, **319**, 70–73.

Marqusee, S. and Baldwin, R. L. (1987) Helix stabilization by Glu–Lys salt bridges in short peptides of *de novo* design. *Proc. Natl. Acad. Sci. USA*, **84**, 8898–8902.

Matsumura, M., Yasumura, S. and Aiba, S. (1986) Cumulative effect of intragenic amino acid replacements on the thermostability of a protein. *Nature (London)*, **323**, 356–358.

Matsumura, M., Yahanda, S. and Aiba, S. (1989a) Site-directed mutagenesis: role of tyrosine 80 in thermal stabilization of kanamycine nucleotidyltransferase. *Eur. J. Biochem.*, **171**, 715–720.

Matsumura, M., Becktel, W. J. and Matthews, B. W. (1988b) Hydrophobic stabilization in T4 lysozyme determined directly by multiple substitutions of Ile 3. *Nature (London)*, **334**, 406–410.

Matsumura, M., Becktel, W. J., Levitt, M. and Matthews, B. W. (1989) Stabilization of phage T4 lysozyme by engineered disulfide bonds. *Proc. Natl. Acad. Sci. USA*, **86**, 6562–6566.

Matthews, B. W., Nicholson, H. and Becktel, W. J. (1987) Enhanced protein thermostability from site-directed mutations that decrease the entropy of unfolding. *Proc. Natl. Acad. Sci. USA*, **84**, 6663–6667.

Matthews, C. R. and Hurle, M. R. (1987) Mutant sequences as probes of protein folding mechanisms. *Bioassays*, **6**, 254–257.

Matthews, C. R. and Westmoreland, D. G. (1975) Nuclear magnetic resonance studies of residual structure in thermally unfolded ribonuclease A. *Biochemistry*, **14**, 4532–4538.

Matthews, C. R., Crisanti, M. M., Gepner, G. L., Velicelebi, G. and Sturtevant, J. M. (1980) Effect of a single amino acid substitution on the thermal stability of the α subunit of tryptophan synthase. *Biochemistry*, **19**, 1290–1293.

Matthews, C. R., Crisanti, M. M., Manz, J. T. and Gepner, G. L. (1983) Effect of a single amino acid substitution on the folding of the α subunit of typtophan synthase. *Biochemistry*, **22**, 1445–1452.

Menendez-Arias, L. and Argos, P. (1989) Engineering protein thermal stability. *J. Mol. Biol.*, **206**, 397–406.

Miller, J. H. (1984) Genetic studies of the *lac* repressor XI. Amino acid replacements in the DNA binding domain of the *Escherichia coli lac* repressor. *J. Mol. Biol.*, **180**, 205–212.

Miller, J. H. and Schmeissner, V. (1979) Genetic studies of the *lac* repressor X. Analysis of missense mutations in the *lacI* gene. *J. Mol. Biol.*, **131**, 223–248.

Miller, J. H., Coulondre, C., Hofer, M., Schmeissner, V., Sommer, H., Schmitz, A. and Lu, P. (1979) Genetic studies of the *lac* repressor IX. Generation of altered proteins by the suppression of nonsense mutations. *J. Mol. Biol.*, **131**, 191–222.

Miller, S., Lesk, A. M., Janin, J. and Chothia, C. (1987) The accessible surface area and stability of oligomeric proteins. *Nature (London)*, **328**, 834–836.

Mitchinson, C. and Baldwin, R. L. (1986) The design and production of semisynthetic ribonucleases with increased thermostability by incorporation of S-peptide analogues with enhanced helical stability. *Proteins Struct. Funct. Genet.*, **1**, 23–33.

Mitchinson, C. and Wells, J. A. (1989) Protein engineering of disulfide bonds in subtilisin BPN'. *Biochemistry*, **28**, 4807–4815.

Nakajima, A. and Scheraga, H. A. (1961) Thermodynamic study of shrinkage and of phase equilibrium under stress in films made from ribonuclease. *J. Am. Chem. Soc.*, **83**, 1575–1584.

Narayana, S. V. L. and Argos, P. (1984) Residue contacts in protein structures and implications for protein folding. *Int. J. Peptides*, **24**, 25–39.

Nemethy, G., Leach, S. J. and Scheraga, H. A. (1966) The influence of amino acid side chains on the free energy of helix–coil transition. *J. Phys. Chem.*, **70**, 998–1004.

Nozaki, Y. and Tanford, C. (1971) The solubility of amino acids and two glycine peptides in aqueous ethanol and dioxane solutions. *J. Biol. Chem.*, **246**, 2211–2217.

Pabo, C. O. and Suchaneck, E. G. (1986) Computer-aided model-building strategies for protein design. *Biochemistry*, **25**, 5987–5991.

Pace, C. N. (1975) The stability of globular proteins. *CRC Crit. Rev. Biochem.*, **3**, 1–43.

Page, M. I. (1984) The energetics and specificity of enzyme–substrate interactions. In *The chemistry of enzyme action* (Page, M. I., ed.), pp. 1–54. Elsevier, Amsterdam.

Page, M. I. and Jencks, W. P. (1971) Entropic contributions to rate accelerations in enzymic and intramolecular reactions and the chelate effect. *Proc. Natl. Acad. Sci. USA*, **68**, 1678–1683.

Pakula, A. A., Young, V. B. and Sauer, R. T. (1986) Bacteriophage λ cro mutations: effects on activity and intracellular degradation. *Proc. Natl. Acad. Sci. USA*, **83**, 8829–8833.

Pantoliano, M. W., Ladner, R. C, Bryan, P. N., Rollence, M. L., Wood, J. F. and Poulos, T. L. (1987) Protein engineering of subtilisin BPN': enhanced stabilization through the introduction of two cysteines to form a disulfide bond. *Biochemistry*, **26**, 2077–2082.

Perry, L. J. and Wetzel, R. (1984) Disulfide bond engineered into T4 lysozyme: stabilization of the protein toward thermal inactivation. *Science*, **226**, 555–557.

Perutz, M. F. (1978) Electrostatic effects in proteins. *Science*, **201**, 1187–1191.

Perutz, M. F. and Lehmann, H. (1968) Molecular pathology of human haemoglobin. *Nature (London)*, **219**, 902–909.

Perutz, M. F. and Raidt, H. (1975) Stereochemical basis of heat stability in bacterial ferredoxins and in haemoglobin. *Nature (London)*, **255**, 256–259.

Perutz, M. F., Kendrew, J. C. and Watson, H. C. (1965) Structure and function of haemoglobin 2: some relations between polypeptide chain configuration and amino acid sequence. *J. Mol. Biol.*, **13**, 669–678.

Privalov, P. L. (1979) Stability of proteins. *Adv. Protein. Chem.*, **33**, 167–241.

Rashin, A. A. and Honig, B. (1984) On the environment of ionizable groups in globular proteins. *J. Mol. Biol.*, **173**, 515–521.

Richards, F. M. (1977) Areas, volumes, packing, and protein stability. *Annu. Rev. Biophys. Bioeng.*, **6**, 151–176.

Richardson, J. S. (1981) The anatomy and taxonomy of protein structure. *Adv. Protein Chem.*, **34**, 167–339.

Rose, G. D., Geselowitz, A. R., Lesser, G. J., Lee, R. H. and Zehfus, M. H. (1985) Hydrophobicity of amino acid residues in globular proteins. *Science*, **229**, 834–838.

Russell, S. T. and Warshel, A. (1985) Calculations of electrostatic energies in proteins. *J. Mol. Biol.*, **185**, 389–404.

Schellman, J. A. (1955a) The thermodynamics of urea solution and the heat of formation of the peptide hydrogen bonds. *C.R. Lab. Carlsberg Ser. Chem.*, **29**, 223–229.

Schellman, J. A. (1955b) The stability of hydrogen-bonded peptide structures in aqueous solution. *C.R. Lab. Carlberg Ser. Chem.*, **29**, 230–259.

Schellman, J. A. (1978) Solvent denaturation. *Biopolymers*, **17**, 1305–1322.

Schulz, G. E. and Schirmer, R. H. (1979) *Principles of protein structure*. Springer-Verlag, New York.

Shoemaker, K. R., Kim, P. S., Brems, D. N., Marquees, S., York, E. J., Chaiken, I. M., Stewart, J. M. and Baldwin, R. L. (1985) Nature of the charged group effect on the stability of the C-peptide helix. *Proc. Natl. Acad. Sci. USA*, **82**, 2349–2353.

Shortle, D. and Lin, B. (1985) Genetic analysis of staphylococcal nuclease: identification of three intragenic 'global' suppressors of nuclease-minus mutations. *Genetics*, **110**, 539–555.

Shortle, D. and Meeker, A. K. (1986) Mutant forms of staphylococcal nuclease with altered patterns of guanidine hydrochloride and urea denaturation. *Proteins Struct. Funct. Genet.*, **1**, 81–89.

Shortle, D., Meeker, A. K. and Freire, E. (1988) Stability mutants of staphylococcal nuclease: large compensating enthalpy–entropy changes for the reversible denaturation reaction. *Biochemistry*, **27**, 4761–4768.

Smith, J. L., Hendrickson, W. A., Honzatko, R. B. and Sheriff, S. (1986) Structural heterogeneity in protein crystals. *Biochemistry*, **25**, 5018–5027.

Stahl, N. and Jencks, W. P. (1986) Hydrogen bonding between solutes in aqueous solution. *J. Am. Chem. Soc.*, **108**, 4196–4205.

Streisinger, G., Mukai, F., Dreyer, W. J., Miller, B. and Horiuchi, S. (1961) Mutations affecting the lysozyme of phage T4. *Cold Spring Harbor Symp. Quant. Biol.*, **26**, 25–30.

Sturtevant, J. M. (1977) Heat capacity and entropy changes in processes involving proteins. *Proc. Natl. Acad. Sci. USA*, **74**, 2236–2240.

Sturtevant, J. M. (1987) Biochemical applications of differential scanning calorimetry. *Annu. Rev., Phys. Chem.*, **38**, 463–488.

Susi, H., Timasheff, S. N. and Ard, J. S. (1964) Near infrared investigation of interamide hydrogen bonding in aqueous solution. *J. Biol. Chem.*, **239**, 3051–3054.

Svensson, L. A., Sjolin, L., Gilliland, G. L. Finzel, B. C. and Wlodawer, A. (1986) Multiple conformations of amino acid residues in ribonuclease A. *Proteins Struct. Funct. Genet.*, **1**, 370–375.

Tanford, C. (1962) Contribution of hydrophobic interactions to the stability of the globular conformation of proteins. *J. Am. Chem. Soc.*, **84**, 4240–4247.

Tanford, C. (1968) Protein denaturation. *Adv. Protein Chem.*, **23**, 121–282.

Tanford, C. (1970) Protein denaturation. Part C. Theoretical models for the mechanism of denaturation. *Adv. Protein Chem.*, **24**, 1–95.

Tanford, C. (1980) *The hydrophobic effect: Formation of micelle and biological membranes*. Wiley, New York.

Thornton, J. M. (1982) Electrostatic interactions in proteins. *Nature (London)*, **295**, 13–14.

Ueda, Y. and Go, N. (1976) Theory of large-amplitude conformational fluctuations in native globular proteins. *Int. J. Peptides*, **8**, 551–558.

Villafranca, J. E., Howell, E. E., Oatley, S. J., Xuong, N.-H. and Kraut, J. (1987) An engineered disulfide bond in dihydrofolate reductase. *Biochemistry*, **26**, 2182–2189.

Wada, A. and Nakamura, H. (1981) Nature of the charge distribution in proteins. *Nature (London)*, **293**, 757–758.

Wagner, G., Kalb, A. J. and Wuthrich, K. (1979) Conformational studies by ^{1}H nuclear magnetic resonance of the basic pancreatic trypsin inhibitor after reduction of the disulfide bond between Cys-14 and Cys-38. *Eur. J. Biochem.*, **95**, 249–253.

Warshel, A. (1987) What about protein polarity? *Nature (London)*, **330**, 15–16.

Wells, J. A. and Powers, D. B. (1986) *In vivo* formation and stability of engineered disulfide bonds in subtilisin. *J. Biol. Chem.*, **261**, 6564–6570.

Wetzel, R., Perry, L. J., Baase, W. A. and Becktel, W. J. (1988) Disulfide bonds and thermal stability in T4 phage lysozyme. *Proc. Natl. Acad. Sci. USA*, **85**, 401–405.

Wolfenden, R., Anderson, L., Cullis, P. M. and Southgate, C. C. B. (1981) Affinities of amino acid side chains for solvent water. *Biochemistry*, **20**, 849–855.

Yutani, K., Ogasawara, K., Aoki, K., Kakuno, T. and Sugino, Y. (1984) Effect of amino acid residues on conformational stability in eight mutant proteins variously substituted at a unique position of the trp synthase α-subunit. *J. Biol. Chem.*, **259**, 14076–14081.

Yutani, K., Ogasawara, K., Tsujita, T. and Sugino, T. (1987) Dependence of conformational stability on hydrophobicity of the amino acid residue in a series of variant proteins substituted at a unique position of tryptophan synthease α subunit. *Proc. Natl. Acad. Sci. USA*, **84**, 4441–4444.

Zales, S. E. and Klibanov, A. M. (1986) Why does ribonuclease irreversibly inactivate at high temperatures? *Biochemistry*, **25**, 5432–5444.

Zipp, A. and Kauzmann, W. (1973) Pressure denaturation of metmyoglobin. *Biochemistry*, **12**, 4217–4228.

4
Stable proteins

4.1 USE OF STABLE PROTEINS

Enzymes are now widely employed in various fields of biotechnology such as in fine organic synthesis and analysis and in manufacturing medicines, pharmaceuticals and others (Katchalski-Katzir and Freeman, 1982; Klibanov, 1983; Wiseman, 1985; Wiseman, 1977–1985; Godfrey and Reichelt, 1983; Arbige and Pitcher, 1989). Mesophilic proteins, with the exception of a few (Kristjansson, 1989), are not stable even at room temperature, because of their marginal stability when isolated from their natural environment (Privalov, 1979). On exposing at higher temperatures the proteins are much more rapidly denatured. More stable enzymes, therefore, have been requested in the fields of biotechnology, and until today much effort has been paid to stabilizing mesophilic enzymes chemically and physically (Martinek *et al.*, 1977; Klibanov, 1979, 1983; Schmid, 1979; Matinek *et al.*, 1980; Mozhaev and Martinek, 1984). As described in Chapter 6, biological stabilization of enzymes has also been recently developed (Ulmer, 1983).

Since the late 1960s the molecular mechanisms of thermostability of enzymes from thermophilic microorganisms have attracted much interest from scientists from the fundamental point of view and from engineers engaging in biotechnology from the applied point of view (Mozhaev and Martinek, 1984; Brock, 1967, 1978, 1985, 1986; Singleton and Amelunxen, 1973; Zuber, 1976; Heinrich, 1976; Kushner, 1978; Shilo, 1978; Friedman, 1978; Amelunxen and Murdock, 1978; Ljungdahl, 1979; Jaenicke, 1981; Mozhaev *et al.*, 1988a; Nosoh and Sekiguchi, 1988, 1990). In the scientific field, interest is focussed on why enzymes from thermophiles are generally more thermostable than their counterparts from mesophiles. The answers to this question—the mechanism of additional stability of thermophilic proteins—will greatly contribute to understanding the structure–stability relationship of proteins, which is one of the attractive topics in life science (see Chapter 3), and to the stabilization of commercially available enzyme proteins (Martinek *et al.*, 1977, 1980; Klibanov, 1979; Daniel *et al.*, 1981; Kristjansson, 1989).

4.2 THERMOPHILIC BACTERIA

Microorganisms, especially bacteria, are able to grow over a wide temperature range: below 0°C to 100°C. The range for growth of each bacterial species, however, is limited, and bacteria are divided into at least three groups depending on their cardinal temperatures, i.e. their minimum, optimum and maximum growth temperatures. Stainer *et al.* (1970) list bacteria with growth spans from $-5°C$ to 22°C as *psychrophiles*, from 10° to 47°C as *mesophiles*, and between 40° and 80°C as *thermophiles*. This classification is not strict, and many bacteria would fall on a borderline and the cardinal temperature may vary with a change of growth conditions.

Thermophiles (Brock, 1986) can be grouped into two classes: facultative thermophiles, which can grow even at the mesophile growth temperature (37°C) but not above 60°C, and obligate thermophiles, which are able to grow only above 40°C. Obligate thermophiles are further classified into two groups: moderate and extreme thermophiles. The optimum growth temperature of the former is generally around 65°C and that of the latter above 70°C. Typical bacteria belonging to the former group are *Bacillus stearothermophilus*, and latter types include *Thermus aquaticus* and *T. thermophilus*, which can grow even at 75–80°C (Stetter, 1986). They have been employed as the materials for biochemical studies, and also as commercial enzymes (Kristjansson, 1989). Recently several new thermophilic species and genera, including *Pyrodictium*, have been isolated (Stetter, 1986). *Pyrodictium* grows optimally at 105°C and has a maximum growth temperature at 110°C—the highest growth temperature yet observed for any organism (Stetter, 1986). The proteins in *Pyrodictium* must be much more stable than the thermophilic proteins so far studied.

4.3 THERMOPHILIC PROTEINS

Thermophilic bacteria can produce in general thermostable enzymes, as compared to their mesophilic counterparts, because of the higher growth temperatures of thermophiles (Singleton and Amelunxen, 1973; Zuber, 1976; Heinrich, 1976; Kushner, 1978; Shilo, 1978; Friedman, 1978; Amelunxen and Murdock, 1978; Ljungdahl, 1979; Janeicke, 1981; Kristjansson, 1989). It was recently shown that pullulanase from *Clostridium thermohydrosulfuricum* is stable up to 90°C, and that its optimum temperature is also 90°C (Saha and Zeikus, 1989). The enzymes from extreme thermophiles belonging to the genera *Bacillus* and *Thermus* are generally stable and active up to at least 80°C and often up to 100°C (Kelly and Deming, 1988; Ward and Moo-Young, 1988). Some thermophilic proteins are thermostable only in the presence of stabilizing factors, but most of them are intrinsically stable (Mozhaev and Martinek, 1984). Other than their thermostability, thermophilic enzymes appear to be catalytically indistinguishable from their mesophilic counterparts; they have similar reactivity and catalytic sites (Amelunxen and Murdock, 1978; Friedman, 1978; Zuber, 1976). In addition, there is no evidence to suggest that thermostable enzymes have compensated for their higher stability by having higher turnover rates (Amelunxen and Murdock, 1978; Friedman, 1978). Thus, the intrinsic thermostability of thermophile proteins has attracted much biochemical attention.

Because of some advantages, such as thermostability and the lack of a red for external cooling in industrial processes, the use of thermostable, thermophilic enzymes in the fields of biotechnology has been proposed (Curtin, 1985). Actually thermostable enzymes are extensively utilized in industrial processing (Godfrey and Reichelt, 1983; Wasserman, 1984), although there are some disadvantages of the high temperature of thermostable enzymes for industrial purposes; i.e. stabilities of substrates and products, oxygen solubility, difficulty of thermal inactivation of enzymes and high optimum temperature (Kristjansson, 1989). However, many thermostable enzymes on the market have been derived from mesophiles (Kristjansson, 1989), partly because of the problem in supplying thermophilic enzymes on a commercial scale. Recent developments in recombination DNA techniques have made it possible to clone thermophilic enzymes into a mesophilic host to produce thermostable, thermophilic enzymes. Such studies will greatly increase the exploitation of thermophiles in biotechnology. It has also become possible to substitute any amino acid residue(s) in to proteins at will as described in Chapter 6. Protein engineering may enable us to thermostabilize industrial enzymes. Unfortunately, construction of thermostable enzymes using this approach is still empirical, because little is known about the molecular basis of thermostability of enzymes, particularly industrial enzymes.

Thermophilic enzymes have been used as materials for protein chemistry and biotechnology, not only for application purposes but also from fundamental points of view. The mechanisms of thermostability of thermophilic proteins will greatly contribute to dissection of protein stability, and the deduced mechanisms for protein stability may in turn give valuable hints in designing for protein stabilization through protein engineering.

4.4 MECHANISM OF STABILITY OF THERMOPHILIC PROTEINS

The upper limit of thermostability of thermophilic proteins is usually 20–30°C higher than that of their mesophilic counterparts, which corresponds to an increase in protein stability of 5–7 kcal mol^{-1} (Finney et al., 1980). As described in Chapter 3, proteins are stabilized by the stabilizing factors, as reflected by the enthalpy term (ΔH), being superior to the destabilizing factors, as reflected by the entropy term ($-T\Delta S$). Additional stability of thermophilic proteins is then due to either or both increase in stabilizing factors and decrease in destabilizing factors. As in almost all cases, if there is no change in destabilizing factors, e.g. due to disulfide formation, the enhanced stability of thermophilic proteins is caused by the change of ΔH, due to the increase in non-covalent contacts or interactions such as van der Waals contacts, H-bonds electrostatic interactions and hydrophobic contacts (Schulz and Schirmer, 1979). The acquired stability of thermophilic proteins (5–7 kcal mol^{-1}) can be derived from only one or two additional electrostatic interactions (Fersht, 1971, 1987; Perutz, 1978) and several additional H-bonds (Schulz and Schirmer, 1979; Pagae, 1984; Fersht et al., 1985; Fersht, 1987) inside the protein globule and seven to ten additional CH_3—groups in the hydrophobic nucleus of the protein molecule (Tanford, 1980) or disulfide bonds (Pantoliano et al., 1987). Such conformational changes in thermophilic protein molecules will not induce drastic rearrangement of their conformations. Structural changes actually caused by an introduced disulfide bond or a single amino acid replacement through protein engineering are discussed in Chapter 6.

We have now principally two molecular mechanisms for the intrinsic thermostability of thermophilic proteins. The mechanisms have usually been deduced using two different approaches. One is the comparison of protein conformations, such as amino acid composition, amino acid sequence and calculated flexibility between thermophilic and mesophilic proteins. Mechanisms deduced by the procedure are the many amino acid replacements throughout whole protein molecules (see the following reviews: Singleton and Amelunxen, 1973; Zuber, 1976; Heinrich, 1976; Kushner, 1978; Shilo, 1978; Jaenicke, 1981; Mozhaev and Martinek, 1984; Mozhaev et al., 1988a; Nosoh and Sekiguchi, 1988, 1990), and the conformational change in whole or secondary structures of protein molecules (Argos et al., 1979; Vihinen, 1987; Querol and Parrilla, 1987; Menendez-Arias and Argos, 1989). This approach is traditional, but has a drawback, as described below. The second approach is the use of wild-type and mutant proteins of different stabilities, in which a single or a few amino acid are substituted and of which the amino acid sequences or three-dimensional structures are known. Therefore, correlation of amino acid substitution with stability can be made (Perutz and Raidt, 1975; Yutani et al., 1977; Biesecker et al., 1977; Grutter et al., 1979; Walker et al., 1980; Ruegg et al., 1982). This procedure is principally similar to protein engineering, because protein engineering is the strategy for artificially replacing a single amino acid in proteins by site-directed mutagenesis to alter protein stability (Chapter 6).

4.4.1 Replacement of many amino acid residues

When structural comparisons aimed at revealing the mechanism of enhanced stability of thermophilic proteins are made between proteins of different phylogenesis and, especially, of different function, the different amino acid composition and sequence resulting from the different phylogenesis and function will include the amino acid(s) responsible for the increased stability of thermophilic proteins. For example, the amino acid sequences of 3-isopropylmalate dehydrogenases from three bacilli of different thermophily, the same phylogenesis and function exhibited only about 60% homology (Sekiguchi et al., 1986a, 1986b; Imai et al., 1987). However, the statistical comparison of the amino acid sequences of many proteins of different stabilities, even of different phylogenesis and function, may provide some important clues for analyzing the enhanced thermostability of thermophilic proteins (Argos et al., 1979; Vihinen, 1987; Querol and Parrillo, 1987; Menendez-Arias and Argos, 1989).

4.4.1.1 Hydrophobic interactions

As described in Chapter 3, hydrophobic interactions may play an important contribution to folding of a polypeptide chain and to stability of its tertiary structure (Schulz and Schirmer, 1979; Tanford, 1978; Creighton, 1983). It is therefore not surprising that many studies have been made to find a correlation between the stability of thermophilic proteins and contents of hydrophobic amino acids by comparing the amino acid compositions of thermophilic and mesophilic proteins.

Total hydrophobicity

With some thermophilic proteins, the correlation of enhanced thermostability with high hydrophobicity has been reported (Wedler and Hoffmann, 1974; Hasegawa and Imahori, 1976; Nakamura et al., 1978; Walker et al., 1980). A similar conclusion was reached with the change in thermostability of a thermophilic enzyme on chemical modification of Lys residues (Sekiguchi et al., 1978). However, sometimes such correlation between hydrophobicity and stability was not found (Biffen and Williams, 1970), and even the opposite relationship has been reported (Wedler et al., 1976).

Such discrepancy in correlation between hydrophobicity and stability can be explained as follows. When a polypeptide chain folds, hydrophobic amino acid residues have a tendency to locate inside its tertiary structure, because the contact of water with non-polar residues is thermodynamically unfavorable (see Chapter 3). In spite of this rule, a part of the hydrophobic amino acid residues occupy about half the surface area of proteins (Lee and Richards, 1971; Chothia and Janin, 1975; Tanford, 1980; Creighton, 1983). These non-polar residues are often organized as hydrophobic surface clusters (Tanford, 1980), and the clusters are functionally important for the proteins (Srere, 1984; Stellwagen, 1984). Since the non-polar residues on the protein surface are harmful to the stability of proteins, the hydrophilization of surface non-polar residues by chemical modification results in enhancing protein stability (Mozhaev et al., 1988b). Localization of non-polar amino acid residues on protein surface may be one of the reasons for the different conclusions on the correlation between hydrophobicity and stability.

Hydrophobicity of amino acids

Total hydrophobicity of proteins is usually calculated as the sum of the hydrophobicities of the constituted amino acids. The hydrophobicity of each amino acid residue was first obtained as the free energy change of the transfer of amino acid from water to the organic phase (Tanford, 1963; Nozaki and Tanford, 1971). Later, Chothia proposed a statistical scale based on the data obtained with 12 proteins; i.e. the greater the number of amino acid residues inside the protein globule, compared with those located on the surface, the more hydrophobic the amino acid (Chothia, 1976). Another statistical scale was determined mainly by the microenvironment of the amino acids in proteins, based on the data obtained with 21 proteins (Ponnuswamy et al., 1980). The hydrophobicity of the microenvironment is calculated as the sum of the hydrophobicities, using Tanford's scale (Tanford, 1963; Nozaki and Tanford, 1971), of the number of neighboring amino acids in the tertiary structure. Other hydrophobic scales are also proposed (Janin, 1979; Wolfenden et al., 1981; Kyte and Doolittle, 1982; Rose et al., 1985a).

As shown in Table 4.1, the two hydrophobic scales proposed by the amino acid extraction scales (e.g. Tanford, 1963) and by the statistical scales (e.g. Chothia, 1976) are in good agreement; Lys, Arg, Asp, Glu, Asn and Gln are least hydrophobic (Table 4.1). However, there is a marked divergency in the two scales. For example, Trp, the most hydrophobic amino acid according to the transfer model, is ranked between hydrophobic and hydrophilic amino acids, according to the statistical scale. In addition, Tyr and Pro, which are closed to Trp in hydrophobicity according to

Table 4.1 — Hydrophobicity scales of amino acids

Residue	Average hydrophobicity[a]	ΔG^b_{atom}	Buried accessible area[c]	Fractional accessible area loss[d]
Ile	1.38	1.9	23.0	0.88
Phe	1.19	2.3	28.7	0.88
Val	1.08	1.5	23.5	0.86
Leu	1.06	1.9	29.0	0.85
Trp	0.81	2.6	41.7	0.85
Met	0.64	2.4	30.5	0.85
Ala	0.62	0.67	31.5	0.74
Gly	0.48	(0)	25.2	0.72
Cys	0.29	0.38	13.9	0.91
Tyr	0.26	1.6	59.1	0.76
Pro	0.12	1.2	53.7	0.64
Thr	−0.05	0.52	46.0	0.70
Ser	−0.18	0.01	44.2	0.66
His	−0.40	0.64	46.7	0.78
Glu	−0.74	−0.76	72.3	0.62
Asn	−0.78	−0.60	62.2	0.63
Gln	−0.85	−0.22	74.0	0.62
Lys	−1.50	−0.57	110.3	0.52
Arg	−2.53	−2.1	93.8	0.64

[a] An average of five hydrophobicity scales, expressed by free energy change (kcal mol^{-1}) on transfer from apolar solvent to water. The values (Eisenberg et al., 1982, 1984) are calculated from the data of Nozaki and Tanford (1971), Chothia (1976), Janin (1979), van Heijne and Blomberg (1979), and Wolfenden et al., (1981). Non-polar, hydrophobic amino acid residues, such as Phe and Leu, have positive values, while polar, hydrophilic residues, such as Glu and Lys, have negative values.
[b] A calculated, atom-based free energy change of transfer, relative to Gly (Eisenberg and McLachlan, 1986).
[c] Mean solvent accessible area of amino acid residues within folded proteins (Å^2). The values are related to residue hydrophobicity in a linear way, i.e. the more hydrophobic the residues, the more completely buried, on average. The values are those for Gly–X–Gly, because the tripeptide surface area depends on the values of its dihedral angles, which are close to the state in proteins.
[d] The average ratio of the buried accessible area on protein folding to the total area in unfolded proteins of amino acid residues. The data are from Rose et al. (1985) and Dworkin and Rose (1987).

the transfer model, can be described as hydrophilic amino acids in the statistical scales. The incorporation of a polar atom in the amino acid side chain decreases its hydrophobicity by 1 to 1.5 kcal mol^{-1} (Chothia and Janin, 1975; Chothia, 1976). Although the large surface areas (aromatic ring) of Trp and Tyr guarantee their high hydrophobicity, the presence of a polar atom decreases the advantage of their localization inside the protein molecule.

Internal environments

There may exist other possible reasons for the discrepancy between the simple extraction scale and the statistical scales. The extraction scale is based on model experiments; i.e. on the transfer of free amino acids from water to an organic solvent. However, in the interior of real proteins the amino acid residues are in contact with other amino acid residues, which influence their geometry and energy (Manavalan and Ponnuswamy, 1977; Crippen and Viswanadhan, 1985). In addition, the amino acid residue must be packed completely in the protein molecule during its folding (Schulz and Schirmer, 1979; Richards, 1977; Lee and Richards, 1971). Some of the bulky amino acid residues such as Trp and Tyr therefore remain exposed on the surface of the protein molecule. In addition to this, organic solvents are used for estimating the hydrophobic scales of amino acids determined by the extraction model (Tanford, 1963; Nozaki and Tanford, 1971) and others (Guy, 1985). The media in these extraction experiments are not appropriate because organic solvents can form hydrogen bonds with polar atoms of amino acids (Guy, 1985; Wolfenden and Radzicka, 1986).

In spite of these defects inherent in the extraction system, the hydrophobicity scale obtained with the system is very often used for calculating the free energy change on substituting an internal amino acid by protein engineering (Chapter 6).

Aliphatic amino acids

The use of aliphatic amino acid content as a more appropriate measure of protein hydrophobicity was proposed, because the aliphatic amino acid residues are more frequently localized inside the protein globule than on its surface (Ikai, 1980). Ikai determined an aliphatic index of proteins as follows:

$$A = X_A + aX_V + b(X_I + X_L)$$

where X_A, X_V, X_I and X_L are the molar ratios of Ala, Val, Ile and Leu in proteins, respectively, and a and b are the numerical coefficients determined by the size of the amino acids. Fig. 4.1 shows that stable proteins from thermophilic organisms have significantly higher aliphatic indexes than mesophilic proteins.

Internal hydrophobicity

The principle and experimental data described above lead us to the conclusion that the greater the number of hydrophobic residue localized inside the protein molecule and the less the number of residues on the protein surface, the more stable the protein, while its total hydrophobicity can remain the same.

Relating to this, Lee and Richards (1971) introduced the idea of 'water-accessible surface area', and Chothia and Janin (1975) related this idea to the total free energy of hydrophobic interactions in the protein according to the following equation:

$$\Delta G_h = \sigma \Delta Å^2$$

where ΔG_h is the change in the total hydrophobic energy caused by a change in the

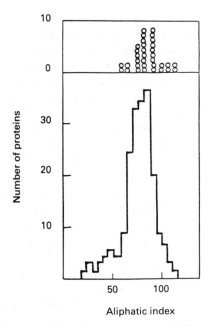

Fig. 4.1—Relationship between aliphatic index and thermostability of proteins. The aliphatic index (see text) is plotted as the abscissa and the number of proteins having a value of aliphatic index as the ordinate. The data are for 34 thermophilic proteins (top part of figure) and 208 mesophilic proteins (bottom part of figure). (Cited from Ikai, 1980.)

accessible surface area equal to $\Delta Å^2$, and σ is the proportionality coefficient equal to 25 kcal mol^{-1} Å. This implies that a decrease in the surface area of a protein responsible for hydrophobic contact with water results in stabilizing the protein. The relation of the water-accessible surface area to stability of proteins was observed when comparing mesophilic and thermophilic proteins (Argos et al., 1979; Zuber, 1981; Stellwagen, 1978; Stellwagen and Wilgus, 1978).

External hydrophobicity

Many reports have been presented on a higher content of Arg and a lower content of Lys in thermophilic proteins (Crabb et al., 1977; Barnes and Stellwagen, 1973; Nakamura et al., 1978; Suzuki and Imahori, 1973; O'brien et al., 1976; Amelunxen and Singleton, 1976; Singleton et al., 1969; Ljungdahl et al., 1976; Kagawa et al., 1976; Frank et al., 1976).

Most of the Arg and Lys residues are located on the surface of proteins (Schulz and Schirmer, 1979; see also Chapter 3), because of their hydrophilic properties due to guanidine and amino groups in the residues, respectively. However, both residues possess relatively longer hydrocarbon chains whose contact with water is thermodynamically unfavourable. The hydrocarbon chain in the Arg residue is one CH_2 group shorter than the Lys residue, and a large guanidine group will provide for a better screening of the hydrocarbon chain than the small amino group. These may partly explain a high content of Arg and a lower content of Lys in thermophilic

4.4 Mechanism of Stability of Thermophilic Proteins

proteins. As described in Chapter 5, the chemical modification, amidination or guanidination, of Lys residues of some proteins increased their stability (Tuengler and Pfleiderer, 1977; Minotani et al., 1979; Cupo et al., 1980). This modification converts Lys residue to arginine-like residues. The enhanced stability of the guanidinated proteins may be due to the screening effect of introduced guanidine group. Hydrophilization of the protein surface accompanied by enhanced stability (Mozhaev et al., 1988b) is also based on the principle; hydrophobicity of a protein surface destabilizes the protein, and hence decrease in hydrophobicity of the protein surface increases the stability of the protein.

4.4.1.2 Compact packing or rigidity

Compact packing

Proteins appear to be densely packed, and the compactness of the interior of proteins is similar to that of crystals (Schulz and Schirmer, 1989). However, there are cavities in the protein molecule; about 25% of the volume of the protein molecule remains unfilled (Chothia, 1984). In the cavities are entrapped small polar molecules such as water (Tilton et al., 1984) which will destabilize proteins through their unfavorable contacts with the hydrophobic core. Therefore, the stability of proteins will increase as protein structure becomes more compact with simultaneous exclusion of water molecules from the cavities.

Lactate dehydrogenase was stabilized on guanidination (Minotani et al., 1979). The hydrogen–deuterium exchange experiments showed that the increased stability is partly related to more compacted conformation on Lys modification (Abe et al., 1983).

The statistical comparison of amino acid sequence of many proteins of different stability showed that favorite amino acid replacements from mesophile to thermophile include replacement with more bulky amino acid residues, which will cause more compact protein structure (Argos et al., 1979; Menendez-Arias and Argos, 1989).

This concept of compact packing will be usefully applied to stabilizing proteins through protein engineering.

Rigidity

Vihinen (1987) proposed that thermostability of proteins arises from the simultaneous effect of several forces, which lead to decreased flexibility (increased rigidity) of protein conformation. Flexibility indices were derived from normalized B-values (B is the atomic temperature factor; Ringe and Petsko, 1986) of individual amino acids in several X-ray crystal structures. Comparison of the flexibility indices and thermostability between six different proteins from many organisms showed that overall flexibility is reduced when thermostability is increased. The relationship between flexibility and thermostability of proteins was discussed by Zuber (1981). Protein molecules require both flexibility and rigidity to function, but the higher the temperature optimum and stability the more rigid is the structure needed to compensate for increased thermal fluctuations. Flexibility of proteins exhibiting the same catalytic activity seems to be about the same at their temperature optimum, but the more rigid thermostable proteins reach the flexibility of thermolabile proteins at higher temperatures.

4.4.1.3 *Electrostatic interactions*

Although the energy of dipole–dipole and dispersion interactions is very small, the number of such interactions in proteins is huge (Schulz and Schirmer, 1979). The contribution of these interactions to protein stability has been theoretically discussed (Gilson *et al.*, 1985; Russell and Warshel, 1985). On the other hand, the number of salt bridges, generally called electrostatic interaction, in proteins is small, but contributes to protein stability by 3–6 kcal mol^{-1} when localized inside the molecule (Fersht, 1971, 1987; Fersht *et al.*, 1985) and by 1–2 kcal mol^{-1} when situated on its surface (Brown *et al.*, 1978; Schulz and Schirmer, 1979; Perutz, 1978; Alber *et al.*, 1987).

The amino acid sequences of ferredoxins from four mesophilic and two thermophilic bacteria exhibit a high homology. Perutz and Raidt (1975) constructed the tertiary structures of five proteins by using the X-ray crystallographic structure of *Micrococcus aerogenes* ferredoxin as a model protein. From the correlation of thermostability to the constructed tertiary structures of proteins, it was suggested that additional electrostatic interactions contribute to the enhanced thermostability of thermophilic ferredoxin.

Recently, the effects of engineering electrostatic interaction on stability were examined with T4 lysozyme (Nicholson *et al.*, 1988) and subtilisin (Erwin *et al.*, 1990).

4.4.1.4 *Hydorgen bonds*

A greater number of hydrogen bonds has been reported for some thermophilic bacteria as compared to their mesophilic counterparts (Boccu *et al.*, 1976; Barnes and Stellwagen, 1973). The numbers of α-helices and β-sheets, in which most H-bonds in protein molecule are involved, however, do not correlate with the thermophily of enzymes (Hachimori *et al.*, 1974; Hibino *et al.*, 1974; Sundaram *et al.*, 1980). Moreover, the correlation between the stability and contents of the H-bond and of secondary structure in proteins has not been established.

α-Helix

Recent investigations, however, suggested a strong correlation between amino acid substitution in the α-helical region and protein stability.

Argos *et al.* (1979) first compared statistically 15 amino acid sequences of mesophilic and thermophilic molecules from three protein families. Later, Menendez-Arias and Argos (1989) compared the amino acid sequence of proteins of different thermostability in more detail, using many more proteins (about 70) of mesophilic and thermophilic molecules from six different protein families form 35 different organisms.

Table 4.2 shows the top ten amino acid replacements from mesophilic to thermophilic proteins (Menendez-Arias and Argos, 1989). The most frequent replacement is Lys to Arg. Since a three-dimensional structure was known for at least one of the amino acid sequences in each protein family, the analysis of preferred

4.4 Mechanism of Stability of Thermophilic Proteins

Table 4.2—Favorable amino acid replacements from mesophilic to thermophilic proteins

Mesophile to[a] thermophile	Ratio of[b] observed/expected values			Preferred locations of the exchange	Effect of the substitution[c]	
	Helix	Sheet	Coil		Hydrophobicity	Flexibility
1. Lys to Arg	1.50	0.56	0.83	Helix	+	−
2. Ser to Ala	1.84	0.73	0.54	Helix	+	−
3. Gly to Ala	1.35	1.45	0.60	Helix & sheet	+	−
4. Ser to Thr	0.42	1.31	1.27	Sheet & coil	+	−
5. Ile to Val	2.45	—	0.41	Helix	−	−
6. Lys to Ala	1.91	1.50	0.20	Helix & sheet	+	−
7. Thr to Ala	1.05	1.87	0.64	Sheet	+	−
8. Lys to Glu	1.14	0	1.29	Coil	+	+
9. Glu to Arg	2.29	0	0.52	Helix	+	−
10. Asp to Arg	1.44	0	1.09	Helix	+	+

[a]Top ten amino acid substitutions.
[b]The ratio is calculated as the percentage of occurrences for the exchange at each type of secondary structure divided by the percentage of amino acids found at each type of secondary structure in all the tertiary structures. In β-sheet, Val to Ile is the preferred replacement.
[c]A (+) represents an increase in the property, while a (−) indicates a decrease. The effect of the substitution on flexibility differs depending on the number of rigid surrounding amino acids. (Cited from Menendez and Argos, 1989.)

residue substitutions which probably achieve thermostability could be examined from a structural context. The major amino acid substitutions were observed only in the α-helical regions, at or next to non-buried residues (Table 4.2). The overall results, which are generally consistent across all the families, suggested that decreased protein flexibility and increased hydrophobicity in α-helical regions are the main stabilizing principles. In addition, it was considered that residues involved in domain interfaces display increased hydrophobicity and decreased flexibility in the thermophiles. The authors emphasized the contribution of their findings to engineering the thermal stability of proteins.

From the comparison of secondary structures of some proteins predicted by the Chou–Fasman method (1978), Querol and Parrilla (1987) proposed that amino acid replacements occur in the amphiphilic helix in external regions of thermophilic proteins, without any change in type and length of secondary structure. The favorite replacements are: Asp to Glu, Lys to Gln, Val to Thr, Ser to Asn, Ile to Thr and Asn to Asp, including Lys to Arg.

4.4.2 Substitution of few amino acids

Besides the mechanism involving many amino acid substitutions for additionally stabilizing proteins, a single or a few amino acid replacements for protein stabilization have been proposed (Mozhaev and Martinek, 1984; Mozhaev et al., 1988a; Nosoh and Sekiguchi, 1988). The proposals have been made by comparing the structures of mesophilic and thermophilic proteins and mostly by comparing the structures of the wild-type and mutant proteins of different stability.

4.4.2.1 Disulfide bonds

The formation of disulfide bonds in a protein decreases the entropy of its unfolded state, thus increasing protein stability. Such a protein-stabilizing effect of the disulfide bond increases as the number of amino acid residues localized in the loop increases (Poland and Scheraga, 1965; Richardson, 1981; Thornton, 1981). Naturally occurring proteins have none or only a few disulfide bonds in their molecules (Schulz and Schirmer, 1979). It is then easy to find whether the correlation of the number of disulfide bond with thermostability exists between mesophilic and thermophilic proteins. Additional intramolecular disulfide bonds have been reported to enhance the stability of some thermophilic enzymes (Wedler et al., 1976; Sundaram et al., 1980; Nakamura et al., 1978).

Recent investigations show that the introduction of disulfide bonds into proteins through proteins engineering can enhance its thermostability, if the bond is appropriately engineered into the interior area of the protein molecule (Chapter 6).

4.4.2.2 Electrostatic interactions

Electrostatic interactions on the protein surface will be easily disrupted by the surrounding water molecules (Schulz and Schirmer, 1979; see also Chapter 3). However, additional interactions on the surface of hemoglobin and ferredoxin have been suggested to contribute to enhanced thermostabilities of the globular proteins (Perutz and Raidt, 1975). Another example for stabilizing thermophilic proteins via additional electrostatic interaction is a thermophilic glyceraldehyde-3-phosphate dehydrogenase (Biesecker et al., 1977; Walker et al., 1980).

Biesecker et al. (1977) compared the three-dimensional structures of D-glyceraldehyde-3-phosphate dehydrogenase from lobster and *Bacillus stearothermophilus*. The apparent conformational difference between them is only three additional electrostatic interactions made by each subunit in the thermophilic enzyme. The interactions were then considered to make a major contribution to the thermostability of the thermophilic enzyme.

The effects of engineered electrostatic interactions on the stability of subtilisin (Erwin et al., 1990) are described in Chapter 6.

4.4.2.3 Hydrophobic interactions

Many studies on the role of hydrophobic interactions in stabilizing proteins have been made by comparing the amino acid compositions of mesophilic and thermophilic proteins (section 4.4.1.1). The comparisons made with wild-type and mutant proteins of different stability also indicate the possible relation of the internal hydrophobicity to the stability of proteins.

α-Subunit of tryptophan synthetase

Yutani et al. (1977) examined the thermostability of the α-subunit of tryptophan synthetase from several E. coli mutants, and found that the replacement of even a single amino acid residue from hydrophilic to hydrophobic residues (for example, Glu to Met at position 49) increases protein stability. The increased stability expressed in free energy change was almost equal to the difference in free energy change between the replaced amino acid residues when transferred from water to ethanol (Yutani et al., 1977), suggesting the correlation between internal hydrophobicity and stability of proteins. Later they confirmed this suggestion by amino acid substitutions at position 49 through protein engineering (Yutani et al., 1984, 1987).

T4 lysozyme

Grutter et al. (1979) examined the three-dimensional structures of the wild-type and temperature-sensitive mutant of T4 phage lysozyme. The mutant strain was generated by use of 2-aminopurine, which induces a single base pair substitution. The mutant protein has a melting temperature about 14°C lower than the wild-type protein. The structural difference between them was only at position 96: Arg in the wild-type and His in the mutant protein; no difference in other regions of the molecule was observed (Fig. 4.2). In the wild-type protein, the side chain of Arg 96 lies on the protein surface between the ring of Tyr 88 and the main chain and C_β of the Leu 91. The hydrocarbon chain of Arg 96 contributes to the major hydrophobic core of the C-terminal lobe of the molecule. Substitution of an imidazole at position 96 presumably destabilizes the hydrophobic core. This provides part of the molecular basis for the decreased stability. The imidazole ring of His 96 lies parallel to the carbonyl oxygen of the same residues, possibly destabilizing the α-helix 82–90.

As will be discussed in Chapter 6, even a single amino acid substitution can increase protein stability by increasing internal hydrophobicity.

4.4.2.4 Hydrogen bonds

Mutation of T4 lysozyme produced a temperature-sensitive protein having a transition temperature 11°C lower than the wild-type protein (Grutter et al., 1987). The difference in stability between the wild-type and mutant proteins corresponds to the change in free energy of 2.9 kcal mol^{-1} at 42°C and at pH 2.0.

The refined X-ray crystal structures of the two proteins showed that Thr 157 is replaced with Ile on mutation, i.e. the replacement of the hydroxyl group (Thr) with an ethyl group (Ile) (see Fig. 6.15). H-bonds to the Thr 155 hydroxyl group are lost in the mutant protein, but tighter binding of a water molecule in the vicinity of the Thr 155 hydroxyl group may partially compensate for this change. The β-carbon of residue 155 shifts ~0.7 Å, and a rotation about the C_α—C_β bond causes a change in the contacts of the side chain γ-methyl group. The side chain of Asp 159 moves ~1.1 Å to accommodate the bulkier ethyl group at position 157.

The main reason for the destabilizing effect of the Thr 157 to Ile substitution appears to be the lack of an H-bond acceptor for the buried main chain amide of Asp 159 in the mutant protein. Even a simple amino acid substitution may alter

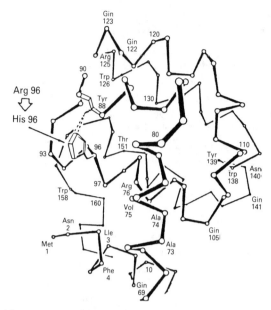

Fig. 4.2—The backbone structure of the C-terminal domain of T4 lysozyme, showing the position of His 96 in the temperature-sensitive mutant. (Cited from Grutter *et al.*, 1979.)

several interactions in the folded state of proteins. The detailed study of many amino acid substitutions in the protien will be described in Chapter 6.

The above example shows the relation of the H-bond to protein stability, although not to protein stabilization.

REFERENCES

Abe, M., Nosoh, Y., Nakanishi, M. and Tsuboi, M. (1983) Hydrogen–deuterium exchange studies on guanidinated pig heart lactate dehydrogenase. *Biochim. Biophys. Acta*, **746**, 176–181.

Alber, T., Sun, D.-P., Wilson, K., Wozniak, J. A., Cook, S. P. and Matthews, B. W. (1987) Contributions of hydrogen bonds of Thr 147 to the thermodynamic stability of phage T4 lysozyme. *Nature (London)*, **330**, 41–46.

Amelunxen, R. E. and Murdock, A. L. (1978) Mechanism of themophily. *CRC Crit. Rev. Microbiol.*, **6**, 343–393.

Amelunxen, R. E. and Singleton, R., Jr (1976) Thermophilic glyceraldehyde-3-phosphate dehydrogenase. In *Enzymes and proteins from thermophilic microorganisms* (H. Zuber, ed.), pp. 107–120. Birkhauser Verlag, Basel.

Arbige, M. V. and Pitcher, W. H. (1989) Industrial enzymology: a look towards the future. *Trends Biotechnol.*, **7**, 330–335.

Argos, P., Rossmann, M. G., Grau, U. M., Zuber, H., Frank, G. and Tratschin, J. D. (1979) Thermal stability and protein structure. *Biochemistry*, **18**, 5698–6703.

Barnes, L. D. and Stellwagen, E. (1973) Enolase from the thermophile *Thermus* X-1. *Biochemistry*, **12**, 1559–1565.

Biesecker, G., Harris, J. I., Thierry, J. C., Walker, J. E. and Wonacott, A. J. (1977) Sequence and structure of D-glyceraldehyde-3-phosphate dehydrogenase from *Bacillus stearothermophilus*. *Nature (London)*, **266**, 328–333.

Biffen, J. H. F. and Williams, R. A. D. (1976) Purification and properties of malate dehydrogenase from *Thermus aquaticus*. In *Enzymes and proteins from thermophilic microorganisms* (H. Zuber, ed.), pp. 157–167. Birkhauser Verlag, Basel.

Boccu, E., Veronese, F. M. and Fontana, A. (1976) Isolation of and some properties of enolase from *Bacillus stearothermophilus*. In *Enzymes and proteins from thermophilic micro-organsims* (H. Zuber, ed.), pp. 229–236. Birkhauser Verlag, Basel.

Brock, T. D. (1967) Life at high temperatures. *Science*, **158**, 1012–1019.

Brock, T. D. (1978) *Thermophilic microorganisms and life at high temperature*. Springer-Verlag, Berlin.

Brock, T. D. (1985) Life at high temperature. *Science*, **230**, 132–138.

Brock, T. D. (ed.) (1986) *Thermophiles: general, molecular, and applied microbiology*. Wiley, New York.

Brown, L. R., DeMarco, A., Richarz, R., Wagner, G. and Wuthrich, K. (1978) The influence of a single salt bridge on static and dynamic features of the globular solution conformation of the bovine pancreatic trypsin inhibitor. *Eur. J. Biochem.*, **88**, 87–95.

Chothia, C. (1976) The nature of the accessible and buried surfaces on proteins. *J. Mol. Biol.*, **105**, 1–12.

Chothia, C. (1984) Principles that determined the structure of proteins. *Annu. Rev. Biochem.*, **53**, 537–572.

Chothia, C. and Janin, J. (1975) Principles of protein–protein recognition. *Nature (London)*, **256**, 705–708.

Chou, P. Y. and Fasman, G. D. (1978) Prediction of the secondary structure of proteins from their amino acid sequences. *Adv. Enzymol.*, **47**, 45–148.

Crabb, J. W., Murdock, A. L. and Amelunxen, R. E. (1977) Purification and characterization of thermolabile glyceraldehyde-3-phosphate dehydrogenase from facultative thermophilic *Bacillus coagulans* KU. *Biochemistry*, **16**, 4840–4847.

Creighton, T. E. (1983) *Proteins.* Freeman, New York.

Crippen, G. M. and Viswanadhan, V. N. (1985) Side chain and backbone potential function for conformational analysis. *Int. J. Peptide Protein Res.*, **25**, 487–509.

Cupo, P., El-Diery, W., Whitney, P. L. and Awad, W. M., Jr (1980) Stabilization of proteins by guanidination. *J. Biol. Chem.*, **255**, 10828–10833.

Curtin, M. E. (1985) The strange but useful archaebacteria. *Biotechnology*, **3**, 36.

Daniel, R. M., Cowan, D. A. and Moran, H. W. (1981) The industrial potential of enzymes from extremely thermophilic bacteria. *Chem. NZ*, **45**, 94–97.

Dworkin, J. E. and Rose, G. D. (1987) Hydrophobicity profiles revisited. In *Methods in protein sequence analysis–1986* (K. A. Walsh, ed.), pp. 573–586. Humana Press, New Jersy.

Eisenberg, D. and McLachlan, A. D. (1986) Solvation energy in protein folding and binding. *Nature (London)*, **319**, 199–203.

Eisenberg, D., Weiss, R. M., Terwilliger, T. C. and Wilcox, W. (1982) Hydrophobic moments and protein structure. *Faraday Symp. Chem. Soc.*, **17**, 109–120.

Eisenberg, D., Schwarz, E., Komaromy, M. and Wall, R. (1984) Analysis of membranes and surface protein sequences with the hydrophobic moment plot. *J. Mol. Biol.*, **179**, 125–142.

Erwin, C. R., Barnett, B. L., Oliver, J. D. and Sullivan, J.F. (1990) Effects of engineered salt bridges on the stability of subtilisin BPN'. *Protein Eng.*, **4**, 87–97.

Fersht, A. R. (1971) Conformational equilibria and the salt bridge in chymotrypsin. *Cold Spring Harbor Symp. Quant. Biol.*, **36**, 71–73.

Fersht, A. R. (1987) The hydrogen bond in molecular recognition. *Trends Biochem. Sci.*, **12**, 301–304.

Fersht, A. R., Shi, J.-P., Knill-Jones, J., Lowe, D. M., Wilkinson, A. J., Blow, D. M., Brick, P., Carter, P., Waye, M. M. Y. and Winter, G. (1985) Hydrogen bonding and biological specificity by protein engineering. *Nature (London)*, **314**, 235–238.

Finney, J. L., Gellathy, B. J., Golton, I. C. and Goodfellow, J. (1980) Solvent effects and polar interactions in the structural stability and dynamics of globular proteins. *Biochem. J.*, **32**, 17–23.

Frank, G., Haberstich, H.-U., Schaer, H. P., Tartshin, J. D. and Zuber, H. (1976) Thermophilic and mesophilic enzymes from *B. caldotenax* and *B. stearothermophilus*:

properties, relationships and formation. In *Enzymes and proteins from thermophilic microorganisms* (H. Zuber, ed.), pp. 375–389. Birkhauser Verlag, Basel.

Friedman, S. M. (ed.) (1978) *Biochemistry of thermophily*. Academic Press, New York.

Gilson, M. K., Rashin, A., Fine, R. and Honing, B. (1985) On the calculation of electrostatic interaction in proteins. *J. Mol. Biol.*, **183**, 503–516.

Godfrey, T. and Reichelt, J. (eds) (1983) *Industrial enzymology*. Nature Press, London.

Grutter, M. G., Hawkes, R. B. and Matthews, B. W. (1979) Molecular basis of thermostability in the lysozyme from bacteriophage T4. *Nature (London)*, **277**, 667–668.

Grutter, M. G., Gray, T. M., Weaver, L. H., Alber, T., Wilson, K. and Matthews, B. W. (1987) Structural studies of mutants of the lysozyme of bacteriophage T4: the temperature-sensitive mutant protein Thr 157 to Ile. *J. Mol. Biol.*, **197**, 315–329.

Guy, H. R. (1985) Amino acid side-chain partition energies and distribution of residues in soluble proteins. *Biophys. J.*, **47**, 61–70.

Hachimori, A., Matsunaga, A., Shimidzu, M., Samejima, Y. and Nosoh, Y. (1974) Purification and properties of glutamine synthetase from *Bacillus stearothermophilus*. *Biochim. Biophys. Acta*, **350**, 461–474.

Hasegawa, A. and Imahori, H. (1976) Studies on α-amylase from a thermophilic bacterium. II. Thermal stability of the thermophilic α-amylase. *J. Biochem. (Tokyo)*, **79**, 469–477.

Heinrich, M. R. (ed.) (1976) *Extreme environments: mechanisms of microbial adaptation*. Academic Press, New York.

Hibino, Y., Nosoh, Y. and Samejima, T. (1974) On the conformation of $NADP^+$-dependent isocitrate dehydrogenase. *J. Biochem. (Tokyo)*, **75**, 553–562.

Ikai, A. (1980) Thermostability and aliphatic index of globular proteins. *J. Biochem. (Tokyo)*, **88**, 1895–1898.

Imai, R., Sekiguchi, T., Nosoh, Y. and Tsuda, K. (1987) The nucleotide sequence of 3-isopropylmalate dehydrogenase gene from *Bacillus subtilis*. *Nucl. Acid Res.*, **15**, 4988.

Jaenicke, R. (1981) Enzymes under extremes of physical conditions. *Annu. Rev. Biophys. Bioeng.*, **10**, 1–67.

Janin, J. (1979) Surface and inside volumes in globular proteins. *Nature (London)*, **277**, 491–492.

Kagawa, Y., Sone, N., Yoshida, M., Hirata, H. and Okamoto, H. (1976) Proton translocating ATPase of a thermophilic bacterium: morphology, subunits, and chemical composition. *J. Biochem. (Tokyo)*, **80**, 141–151.

Katchalski-Katzir, E. and Freeman, A. (1982) Enzyme engineering reaching maturity. *Trends Biochem. Sci.*, **7**, 427–431.

Kelly, R. M. and Deming, J. W. (1988) Extremely thermophilic archaebacteria: biological and engineering considerations. *Biotechnol. Progress*, **4**, 47–62.

Klibanov, A. M. (1979) Enzyme stabilization by immobilization. *Anal. Biochem.*, **93**, 1–25.

Klibanov, A. M. (1983) Immobilized enzymes and cells as practical catalysis. *Science*, **219**, 722–727.

Kristjansson, J. K. (1989) Thermophilic organisms as sources of thermostable enzymes. *Trends Biotech.*, **7**, 349–353.

Kushner, D. J. (ed.) (1978) *Microbial life in extreme environments*. Academic Press, New York.

Kyte, J. and Doolittle, R. (1982) A simple method for displaying the hydrophobic character of a protein. *J. Mol. Biol.*, **157**, 105–132.

Lee, B. and Richards, F. M. (1971) The interpretation of protein structures: estimation of static accessibility, *J. Mol. Biol.*, **55**, 379–400.

Ljungdahl, L. G. (1979) Physiology of thermophilic bacteria. *Adv. Microbiol.*, **19**, 149–243.

Ljungdahl, L. G., Sherod, D. W., Moore, M. R. and Andreesen, J. R. (1976) Properties of enzymes from *Clostridium thermoaceticum* and *Clostridium formicoaceticum*. In *Enzymes and proteins from thermophilic microorganisms* (H. Zuber, ed.), pp. 237–248. Birkhauser Verlag, Basel.

Manavalan, P. and Ponnuswamy, P. K. (1977) A study of the preferred environment of amino acid residues in globular proteins. *Arch. Biophys. Chem.*, **184**, 476–487.

Martinek, K., Klibanov, A. M. and Berezin, I. V. (1977) General principles of enzyme stabilization. *J. Solid-Phase Biochem.*, **2**, 343–385.

Martinek, K., Mozhaev, V. V. and Berzin, I. V. (1980) General principle of stabilization and reactivation of enzymes. In *Enzyme engineering: future directions* (L. B. Wingard, I. V. Berezin and A. A. Klyosov, eds), Ch. 1. Plenum Press, New York.

Matsumura, M., Yahanda, S., Yasumura, S., Yutani, K. and Aiba, S. (1988) Role of tyrosine-80 in the stability of kanamycin nucleotidyltransferase analyzed by site-directed mutagenesis. *Eur. J. Biochem.*, **171**, 715–720.

Menendez-Arias, L. and Argos, P. (1989) Engineering protein thermal stability: sequence statistics point to residues substitutions in α-helices. *J. Mol. Biol.*, **206**, 397–406.

Minotani, N, Sekiguchi, T., Bautista, J. G. and Nosoh, Y. (1979) Basis of thermostability in pig heart lactate dehydrogenase treated with O-methylisourea. *Biochim. Biophys. Acta*, **581**, 334–341.

Mozhaev, V. V. and Martinek, K. (1984) Structure–stability relationship in proteins: new approaches to stabilizing enzymes. *Enzyme Microb. Technol.*, **6**, 50–59.

Mozhaev, V. V., Berezin, I. V. and Martinek, K. M. (1988a) Structure–stability relationship in proteins: fundamental task and strategy for the development of stabilized enzyme catalysts for biotechnology. *CRC Crit. Rev. Bochem.*, **23**, 235–282.

Mozhaev, V. V., Siksnis, V. A., Melik-Nubarov, N. S., Galkantrite, N. Z., Denis, G. J., Butkus, E. P., Zaslevsky, B. Yu., Mestechkina, N. M. and Martinek, K. (1988b) Protein stabilization via hydrophilization: covalent modification of trypsin and α-chymotrypsin. *Eur. J. Biochem.*, **173**, 147–154.

Nakamura, S., Ohta, S., Arai, K., Oshima, T. and Kajiro, K. (1978) Studies of polypetide-chain-elongation factors from an extreme thermophile, *Thermus thermophilus*. *Eur. J. Biochem.*, **92**, 533–543.

Nicholson, H., Becktel, W. J. and Matthews, B. W. (1988) Enhanced protein thermostability from designed mutations that interact with α-helical dipoles. *Nature (London)*, **336**, 651–656.

Nosoh, Y. and Sekiguchi, T. (1988) Protein thermostability: mechanism and control through protein engineering. *Biocatalysis*, **1**, 257–273.

Nosoh, Y. and Sekiguchi, T. (1990) Protein engineering for thermostability. *Trends Biotech.*, **8**, 16–20.

Nozaki, Y. and Tanford, C. (1971) Solubility of amino acids and two glycine peptides in aqueous ethanol and dioxane solutions: establishment of a hydrophobicity scale. *J. Biol. Chem.*, **246**, 2211–2217.

O'brien, W. E., Brewer, J. M. and Ljungdahl, L. G. (1976) Chemical, physical and enzymatic comparison of formyltetrahydrofolate synthetase. In *Enzymes and proteins from thermophilic microorganisms* (H. Zuber, ed.), pp. 249–262. Birk hauser Verlag, Basel.

Page, M. I. (1984) The energetics and specificity of enzyme–substrate interaction. In *The Chemistry of Enzyme Action* (M. I. Page, ed.), pp. 1–54. Elsevier, Amsterdam.

Pantoliano, M. W., Ladner, R. C., Bryan, P. N., Rollence, M. L., Wood, J. F. and Poulos, T. L. (1987) Protein engineering of subtilisin BPN': enhanced stabilization through the introduction of two cysteines to form a disulfide bond. *Biochemistry*, **26**, 2077–2082.

Perutz, M. F. (1978) Electrostatic effects in proteins. *Science*, **201**, 1187–1191.

Perutz, M. F. and Raidt, H. (1975) Stereochemical basis of heat stability in bacterial ferredoxins and haemoglobin A2. *Nature (London)*, **255**, 256–259.

Poland, D. C. and Scheraga, H. A. (1965) Statistical mechanics of noncovalent bonds in polyamino acids. VII. Covalent loops in proteins. *Biopolymers*, **3**, 379–399.

Ponnuswamy, P. K., Prebhakaran, M. and Manavalan, P. (1980) Hydrophobic packing and spatial arrangement of amino acid residues in globular proteins. *Biochim. Biophys. Acta*, **623**, 301–316.

Privalov, P. L. (1979) Stability of proteins. *Adv. Protein Chem.*, **33**, 167–241.

Querol, E. and Parrilla, A. (1987) Tentative rules for increasing the thermostability of enzymes by protein engineering. *Enzyme Microbiol. Technol.*, **9**, 238–244.

Richards, F. M. (1977) Areas, volume, packing and protein structure. *Annu. Rev. Biophys. Bioeng.*, **6**, 151–176.

Richardson, J. S. (1981) The anatomy and taxonomy of protein structure. *Adv. Protein Chem.*, **34**, 167–339.

Ringe, D. and Petsko, G. A. (1986) Study of protein dynamics by X-ray diffraction. *Methods Enzymol.*, **131**, 389–447.

Rose, G. D., Gaselowitz, A., Lesser, G., Lee, R. and Zehfus, M. (1985a) Hydrophobicity of amino acid residues in globular proteins. *Science*, **229**, 834–838.

Rose, G. D., Gierasch, L. M. and Smith, J. A. (1985b) Turns in peptides and proteins. *Adv. Protein Chem.*, **37**, 1–109.

Ruegg, C., Ammer, D. and Lerch, K. (1982) Comparison of amino acid sequence and thermostability of tyrosinase from three wild type strains of *Neurospora crassa*. *J. Biol. Chem.*, **257**, 6420–6426.

Russell, S. T. and Warshel, A. (1985) Calculation of electrostatic energies in proteins: the energetics of ionized groups in bovine pancreatic trypsin inhibitor. *J. Mol. Biol.*, **185**, 389–404.

Saha, B. C. and Zeikus, J. G. (1989) Novel highly thermostable pullulanase from thermophiles. *Trends Biotechnol.*, **7**, 234–239.

Schmid, R. D. (1979) Stabilized soluble enzymes. *Adv. Biochem. Eng.*, **12**, 41–118.

Schulz, G. E. and Schirmer, R. D. (1979) *Principles of protein structure*. Springer-Verlag, Berlin.

Sekiguchi, T., Matsunaga, A., Shimamura, A., Nosoh, Y., Hachimori, A. and Samejima, T. (1978) On the thermostability of amino acid modified glutamine synthetase from *Bacillus stearothermophilus*. In *Biochemistry of thermophily* (S.M. Freidman, ed.) pp. 345–357. Academic Press, New York.

Sekiguchi, T., Ortega-Cesena, J., Nosoh, Y., Ohashi, S., Tsuda, K. and Kanaya, S. (1986a) DNA and amino-acid sequences of 3-isopropylmalate dehydrogenase of *Bacillus coagulans*: comparison with the enzymes of *Saccharomyces cerevisiae* and *Thermus thermophilus*. *Biochim. Biophys. Acta*, **867**, 36–44.

Sekiguchi, T., Suda, M, Ishii, T., Nosoh, Y. and Tsuda, K. (1986b) The nucleotide-sequence of 3-isopropylmalate dehydrogenase from *Bacillus caldotenax*. *Nucl. Acid Res.*, **15**, 853.

Shilo, M. (ed.) (1978) *Strategies of microbial life in extreme environments*. Chemie-Verlag, Weinheim.

Singleton, R., Jr and Amelunxen, R. E. (1973) Proteins from thermophilic microorganisms. *Bacteriol. Rev.*, **37**, 320–342.

Singleton, R., Jr, Kimmel, J. R. and Amelunxen, R. E. (1969) Amino acid composition and other properties of thermostable glyceraldehyde-3-phosphate dehydrogenase from *Bacillus stearothermophilus*. *J. Biol. Chem.*, **244**, 1623–1630.

Srere, P. A. (1984) Why are enzyme so big? *Trends Biochem. Sci.*, **9**, 387–390.

Stainer, R. Y., Kostiw, L. L. and Adelberg, E. A. (1970) *The microbial world*, 3rd edn. Prentice Hall, Englewood cliffs, NJ.

Stellwagen, E. (1978) Relationship of protein thermostability to accessible surface area. *Nature (London)*, **275**, 342–343.

Stellwagen, E. (1984) Strategies for increasing the stability of enzymes. *Ann. NY Acad. Sci.*, **434**, 1–6.

Stellwagen, E. and Wilgus, H. (1978) Thermostability of proteins. In *Biochemistry of thermophily* (S. M. Friedman, ed.), pp. 223–232. Academic Press, New York.

Stetter, K. O. (1986) Diversity of extremely thermophilic *Archaebacteria*. In *Thermophiles: general, molecular, and applied microbiology* (T. D. Brock, ed.), pp. 39–74. Wiley, New York.

Sundaram, T. K., Chell, R. M. and Wilkinson, A. E. (1980) Monomeric malate dehydrogenase from a thermophilic *Bacillus*: molecular and kinetic characteristics. *Arch. Biochem. Biophys.*, **199**, 515–525.

Suzuki, K. and Imahori, K. (1973) Glyceraldehyde 3-phosphate dehydrogenase of *Bacillus stearothermophilus*: kinetics and physicochemical studies. *J. Biochem. (Tokyo)*, **74**, 955–970.

Tanford, C. (1963) Contribution of hydrophobic interactions to the stability of the globular conformation of proteins. *J. Am. Chem. Soc.*, **84**, 4240–4247.

Tanford, C. (1978) The hydrophobic effect and the organization of living matter. *Science*, **200**, 1012–1018.

Tanford, C. (1980). *The hydrophobic effect: formation of micells and biological membranes*. Wiley, New York.

Thornton, J. M. (1981) Disulfide bridges in globalar proteins. *J. Mol. Biol.*, **151**, 261–287.

Tilton, R. F., Jr, Kuntz, I. D. and Petsko, G.A. (1984) Cavities in proteins: structure of a metmyoglobin-xenon complex solved to 1.9 Å, *Biochemistry*, **23**, 2849–2857.

Tuengler, P. and Pfleiderer, G. (1977) Enhanced heat, alkaline, and tryptic stability of acetamidinated pig heart lactate dehydrogenase. *Biochim. Biophys. Acta*, **484**, 1–8.

Ulmer, K. M. (1983) Protein engineering, *Science*, **219**, 666–671.

Van Heijne, G. and Blomberg, C. (1979) Trans-membrane translocation of proteins: the direct transfer model. *Eur. J. Biochem.*, **97**, 175–181.

Vihinen, M. (1987) Relationship of protein flexibility to thermostability. *Protein Eng.*, **1**, 477–480.

Walker, J. E., Wonacott, A. J. and Harris, J. I. (1980) Heat stability of a tetrameric enzyme, D-glyceraldehyde-3-phosphate dehydrogenase. *Eur. J. Biochem.*, **108**, 581–586.

Ward, O. P. and Moo-Young, M. (1988) Thermostable enzymes. *Biotech. Adv.*, **6**, 39–69.

Wasserman, B. P. (1984) Thermostable enzyme production. *Food Technol.*, **38**, 80–89.

Wedler, F. C. and Hoffmann, F. M. (1974) Glutamine synthetase of *Bacillus stearothermophilus*. I. Purification and basic properties. *Biochemistry*, **13**, 3207–3214.

Wedler, F. C., Hoffmann, F. M., Kenney, R. and Carfi, J. (1976) Maintenance of specificity, information, and thermostability in thermophilic *Bacillus sp.* glutamine synthetase. In *Enzymes and proteins from thermophilic microorganisms* (H. Zuber, ed.), pp. 187–197. Birkhauser Verlag, Basel.

Wiseman, A. M. (1985) *Handbook of enzyme biotechnology*, 2nd edn. Halsted, London.

Wiseman, A. M. (ed.) (1977–1985) *Topics in enzyme and fermentaion biotechnology*, Vols 1–11. Wiley, London.

Wolfendon, R. and Radzicka, A. (1986) How hydrophilic is tryptophan? *Trends Biochem. Sci.*, **11**, 69–70.

Wolfenden, R., Andersson, L., Cullis, P. and Southgate, C. (1981) Affinities of amino acid side chains for solvent water. *Biochemistry*, **20**, 849–855.

Yutani, K., Ogasawara, K., Sugino, Y. and Matsushiro, A. (1977) Effect of single amino acid substitution on stability of conformation of a protein. *Nature (London)*, **267**, 274–275.

Yutani, K., Ogawawara, K., Aoki, K., Kakuno, T. and Sugino, Y. (1984) Effect of amino acid residues on conformational stability in eight mutant proteins variously substituted at unique position of the tryptophan synthetase α-subunit. *J. Biol. Chem.*, **259**, 14076–14081.

Yutani, K., Ogasawara, K., Tsujita, T. and Suginio, Y. (1987) Dependence of conformational stability of hydrophobicity of the amino acid residue in a series of variant proteins substituted at a unique position of tryptophan synthetase α-subunit. *Proc. Natl. Acad. Sci. USA*, **84**, 4441–4444.

Zuber, H. (ed.) (1976) *Enzymes and proteins from thermophilic microoganisms*. Birkhauser, Basel.

Zuber, H. (1981) Structure and function of thermophilic enzymes. In *Structural and functional aspects of enzyme catalysis* (H. Eggerer and R. Huber, eds), pp. 114–127. Springer-Verlag, New York.

5

Stabilization through chemical and physical modifications

Today, enzymes and cells or organisms in which enzymes are originally involved are widely employed in biotechnology (Katchalski-Katzir and Freeman, 1982; Bungay and Belfort, 1987); e.g. as biocatalysts for producing organic intermediates, specific chemicals and pharmaceuticals and as a biosensor. Most mesophilic enzymes, with some exceptions, are marginally stabilized (Privalov, 1979), and during a long term of use and store the commercial enzymes from mesophiles decrease and lose their activity. Thus, to obtain much more stable enzymes, many approaches have been made to stabilize commercial enzymes by various procedures such as chemical, physical and biological modifications (Mozhaev and Martinek, 1984; Mozhaev et al., 1988a; Nosoh and Sekiguchi, 1988). Instead of stabilizing mesophilic enzymes, the stable enzymes from thermophilies have been considered for commercial use (Kristjansson, 1989).

Chemical modification of proteins has developed to modify or identify the functional amino acid residue(s) in proteins (Means and Feeney, 1971). The biocatalysts were immobilized from the point of their reuse, through chemical and physical procedures (Klibanov, 1979, 1983). It was shown that enzymes are stabilized by these procedures (Klibanov, 1979, 1983; Mozhaev and Martinek, 1984; Mozhaev et al., 1988a). Recently, biological modification has been shown to alter various properties of proteins by replacing amino acid residue(s) through mutagenesis of the genes coding for proteins, including site-directed mutagenesis (protein engineering) (Ulmer, 1983). Thus protein stability can be modified through this modification strategy.

In this chapter, the chemical and physical modifications of proteins are briefly described, and the procedures, together with the biological modification (Chapter 6), are compared with reference to their applicability to protein stability and stabilization.

5.1 STRATEGIES FOR PROTEIN STABILIZATION

As described above, there exist three different strategies for stabilizing proteins depending on the principles by which proteins are stabilized:

(1) Amino acid residues of proteins are chemically modified, without any carriers, by various specific chemicals. This *chemical modification* produces proteins of varied functions such as enzyme activity (Shaw, 1970; Cohen, 1970; Means and Feeney, 1971; Hirs and Timasheff, 1977; Torchilin and Martinek, 1979; Branner-Yorgensen, 1983; Kaiser *et al.*, 1985; Holmes, 1987). Chemical modification, if applied carefully, can also bestow stable conformations to proteins without any drastic conformational change (e.g. Mozhaev *et al.*, 1988a).

(2) Proteins are bound through chemical bonds to gels or trapped physically or chemically into gels. This *immobilization* technique has been applied to enzymes and cells or organisms, and has been widely employed in biotechnology (Klibanov, 1979, 1983; Mozhaev and Martinek, 1984; Mozhaev *et al.*, 1988a; Mosbach, 1976, 1987).

(3) A single or only a few amino acids in proteins are substituted, especially through site-directed mutagenesis (protein engineering) (Ulmer, 1983), producing mutants of modified functions. Protein engineering started very recently, but nowadays it is considered to be a promising technique for stabilizing proteins, as well as for analyzing protein stability. In contrast to the other chemical and physical techniques the candidate site for protein stabilization can be specifically modified at will by this *biological modification* (see Chapter 6).

Recently, a new technique for stabilizing proteins has been proposed (Shami *et al.*, 1989). This technique is based on the complex formation of enzymes and their specific antibodies. For example, the thermostability of human salivary α-amylase complexed with its specific rabbit polyclonal antibody was much higher than that of free enzyme.

5.1.1 Chemical modification

Chemical modification has developed with the growing needs for indentifying and modifying the functional amino acid residue(s) in proteins. The functional residues are located on or near the surface of proteins (Alberts *et al.*, 1983), and thus are reactive to chemical reagents. One of the pioneer works on chemical modification of proteins is the reaction of the hydroxyl group of the functional Ser residue in proteins, e.g. serine proteases, with diisopropyl fluorophosphate. The reaction proceeds equivalently and completely inactivates the serine proteases, indicating the important role of Ser in catalytic activity of the proteases (Mounter *et al.*, 1963). Since then many chemical modifiers have been developed for protein modification (Kaiser *et al.*, 1985; Lundblad and Noyes, 1985).

The chemically reactive groups of amino acid residues are limited. They are nucleophiles and proton donors or acceptors: carboxyl group (C-terminal, Asp and Glu), imidazolyl group (His), sulfhydryl group (Cys), amino group (N-terminal and Lys), hydroxyl group (Try, Thr and Ser), guanidinium group (Arg), indolyl group (Trp) and methylthio group (Met). None of the hydrophobic side chains are reactive.

Since the functional groups of proteins possess the same reactivity toward a reagent, efforts to modify a certain group by a reagent are usually unsuccessful. Therefore the competition of different functional groups for the same reagent causes non-specific modification. Even for the group-specific reagents, reaction conditions can be chosen under which amino acid residues of only one type are selectively modified. This is achieved, for example, by varying the pH and thus making only one form of the functional group (either protonated or deprotonated) reactive, or reversible blocking of the functional groups (e.g. Nieto and Palacian, 1983). In spite of the defects or limitations of chemical modification, a large number of specific reagents with a high selectivity are now available (Lundblad and Noyes, 1985), because of the great advance in studies on the chemical modification of proteins (Moore and Free, 1985; Lundblad and Noyes, 1985).

Chemical modification has also been applied to stabilize proteins through modification of some amino acid residues (Mozhaev and Martinek, 1984; Mozhaev et al., 1988a). Some examples are described in Section 5.3.

5.1.2 Immobilization

Immobilized biocatalysts are enzymes, cells or organisms, which are in a state that permits their reuse (e.g. Klibanov, 1983). Because of the reuse and stability of biocatalysts and the ease of isolation of reaction products, a number of processes using immobilized enzymes have received wide application in biotechnology and are now well established on an industrial scale (e.g. Chibata et al., 1979; Trevan, 1980; Katchalski-Katzir and Freeman, 1982; Klibanov, 1983; Vandamme, 1983; Wood and Calton, 1984; Jensen et al., 1984; Klein and Langer, 1986; Kricka and Thorpe, 1986).

Immobilized biocatalysts comprise biocatalysts chemically or physically immobilized by attachment to soluble polymers or insoluble carriers or by entrapment in membrane systems (Mosbach, 1976, 1987). Some examples of immobilization are shown in Fig. 5.1. As shown in the figure, the immobilization technique is considered

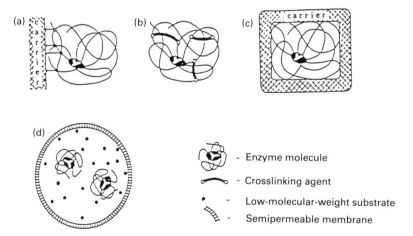

Fig. 5.1—Schematic representation of immobilization of proteins. Chemical immobilization: (a) attachment to a carrier through chemical bonds, and (b) cross-linking a protein by bi- (or poly)-functional agents; physical immobilization (c) entrapping into gels; and (d) microencapsulation of proteins. (Cited from Martinek and Mozhaev, 1985).

to restrict the mobility of protein molecules or their fragments in space (Katchalski-Katzir et al., 1971; Zaborsky, 1973). Thus, immobilized enzymes can be more stable than free enzymes (Klibanov, 1979; Schmid, 1979; Mozhaev and Martinek, 1984). Since immobilization does not usually cause any significant conformational change in proteins, as revealed by physical methods (Martinek and Mozhaev, 1985), the immobilization technique can be applied to fundamental studies in biochemistry such as subunit association or dissociation of proteins (Mosbach, 1976, 1987).

5.2 EVALUATION OF THREE STRATEGIES FOR STABILIZING PROTEINS

There are two reasons for stabilizing proteins. One is the use of stabilized enzymes for producing various chemicals, and another is to provide a sample for analyzing protein stability.

For industrial purposes, stable proteins can be obtained from thermophiles (Chapter 4) without any modification, but enzymes useful for biotechnology are not always produced from thermophiles on a large scale (Kristjansson, 1989). Although some thermophilic enzyme genes can be expressed in mesophiles, none of the genes coding for commercially available enzymes has been successfully expressed in mesophiles on a commercial scale (Kristjansson, 1989). From the applied point of view, immobilization has the highest applicability among the three procedures, as described in a previous section, in the points of reuse and isolation of reaction products. In addition, immobilized enzymes are stable enough for industrial purposes.

Immobilized enzymes, however, cannot be suitable materials for revealing the molecular reasons of protein stability. As discussed in Chapters 3 and 6, the analysis of ΔG for protein unfolding and of protein structure, especially X-ray crystal structure, is essential for quantitavely understanding protein stability. In this respect, immobilized proteins have limitations in the applicability of chemical and physical analytical procedures, because the matrix in or to which proteins are immobilized will strongly interfere with the analysis. For the purpose of dissecting protein stability, then, chemical and biological modifications of proteins appear suitable, because the modifications do not use any carriers which interfere, for example, with ΔG and structure measurements.

Chemical modification has a high potentiality in stabilizing proteins (Mozhaev et al., 1988a), but the technique has some drawbacks:

(1) Modification often causes unfavorable conformational changes and denaturation (Cohen, 1970). The activities and conformations of the native and modified proteins should be examined.

(2) Sometimes undesired side reactions occur, if modification is carried out under rather harsh conditions (at high or low pH, in the presence of oxidants and reductants, etc.). In some cases, such side reactions can be avoided by reversible protection of the functional groups which undergo the side reactions (Mozhaev et al., 1988a).

(3) Chemical modification generally yields several derivatives which differ in the number and position of functional groups modified (Cohen, 1970). It seems difficult to separate each derivative, even by high-performance liquid chromatography.

(4) It is difficult to identify the modified amino acids, because the acid hydrolysis of modified proteins may cause cleavage of the less stable chemical bonds.

On the other hand, biological modification does not have such defects. Through site-directed mutagenesis, for instance, any amino acid, even with a non-polar side chain, at any position in a protein can be replaced by any other amino acid at will, if the gene coding for the protein can be isolated from an organism and can be cloned in microorganisms such as *E. coli* (Chapter 6). This biological modification, however, can only be made based on information about the structure–stability relationship of the protein. As described in Chapters 3 and 6, protein engineering is a promising strategy for analyzing protein stability and also for stabilizing proteins, which are closely correlated.

5.3 STABILIZATION BY CHEMICAL MODIFICATION

In spite of the drawbacks inherent to chemical modification as described above, the procedure has long been used to stabilize proteins, because of its facility in this application (Torchilin and Martinek, 1979; Mozhaev et al., 1988a). In some cases, the mechanism of protein stability, not at the molecular level but at the whole protein conformational level, can be deduced by chemical modification of proteins. In this section, therefore, some examples of stability and stabilization through chemical modification are described.

5.3.1 Possible reasons for altered thermostability

Induced conformational change

Chemical modification such as acetylation of α-amylase (Urabe *et al.*, 1973) or methylation of alcohol dehydrogenase (Tsai *et al.*, 1974) produced thermostable derivatives. The stabilization was considered to be due to some conformational change of proteins, but molecular reasons for such structural changes have not been examined, and the characters of conformational changes induced by the modification could not be predicted.

Modification of key functional residues

Torchilin and Martinek (1979) proposed the presence of *key functional groups* in proteins; i.e. on addition of increasing concentrations of reagents some key functional groups on the protein surface are first modified without any significant change in conformation and stability, and the groups inside the protein globule are then modified, which causes alteration of the balance of intramolecular interaction. Increased thermostability of alkylated α-chymotrypsin was explained by this mechanism (Torchilin *et al.*, 1979).

Key groups of proteins, however, cannot be assessed from the known three-dimensional structure of proteins, and it is difficult to point out key groups of proteins through chemical modification experiments (Torchilin and Martinek, 1979; Mozheav *et al.*, 1988).

Cross-linking

As in the case of disulfide bonds (Chapter 3), cross-linking proteins with bi (or poly)-functional agents increases their stability (Torchilin *et al.*, 1978; Peters and

Richards, 1979; Torchilin and Martinek, 1979; Sekiguchi et al., 1979a, 1979b; DeRenobales and Welch, 1980).The external braces may increase the entropy of the unfolded state of proteins.

Introduction of new H-bond of salt bridge

New functional groups, polar of charged, introduced into proteins by chemical modification may form additional H-bonds or electrostatic interaction. Treatment of lactate dehydrogenase with acetoimidate yielded thermostabilized proteins (Muller, 1981). The increased stability of the enzyme was suggested to be due to a newly formed salt bridge. Only X-ray crystal structure of the modified product may confirm such an additional H-bond or salt bridge. However, the heterogeneity of modified proteins, as mentioned above and discussed later, has made it difficult or impossible to determine their three-dimensional structures.

Hydrophobization

The chemical modification of hydrophilic residues of proteins by hydrophobic reagents (e.g. methylation of Lys) yielded proteins of increased (Shatsky et al., 1973) or decreased stability (Urabe et al., 1978; Kagawa and Nukiwa, 1981). Although hydrophobicity on the surface of proteins destabilizes proteins, many hydrophobic residues on the protein surface very often accumulate to form surface hydrophobic clusters, along with polar and charged amino acid residues (Krigbaum and Komoriya, 1979; Chothia, 1984; Burley and Petsko, 1985). If a residue to be modified is in the vicinity of such a cluster, a modifier with a suitable chain length may come into contact with it, thus increasing the stability of the protein, because of additional hydrophobic interactions (Shatsky et al., 1973).

Hydrophilization

In contrast to the above reason for increased thermostability, the decrease in surface hydrophobicity character (increase of surface hydrophilicity) by chemical modification might result in an increase of protein stability.

Amidination with imidoesters or guanidination with O-methylisourea of amino groups of Lys, both yielding Arg-like residues, are examples of protein stabilization by this procedure (Tuengler and Pfleiderer, 1977; Minotani et al., 1979; Cupo et al., 1980; Muller, 1981; Fojo et al., 1982; Shibuya et al., 1982). The length of the hydrocarbon chain of the Arg residue is shorter than that of the Lys residue, indicating a lesser destabilizing effect of the Arg residue. The guanidinium group of Arg is more hydrophilic and bulkier than the amino group of Lys. The former will more effectively screen the thermodynamically unfavorable contact with water of the hydrocarbon chain than the latter. Thus the conversion of Lys to Arg on the protein surface will decrease the protein destabilising effect of protein hydrophobicity.

Chemical modification of an enzyme with polysaccharides produced thermostabilized derivatives (Marshall, 1978). Two possible reasons for the increased stability can be considered. One is a slight increment of surface hydrophilicity by introduced carbohydrates, the change of hydrophilicity on introduction of sugar being calculated by the scale of Hansch and Leo (1979). However, a more probable reason is the entropic effect of saccharides which perturbate the water structure around the sugars in the native state of enzymes (Arakawa and Timasheff, 1982).

One of the cleanest demonstrations of protein stabilization by enhancing surface hydrophilicity is the amination of Tyr residues on the surface area of trypsin or carboxylation of Lys residues of α-chymotrypsin (Mozhaev et al., 1988b). The amination of the hydroxyl group of Tyr directly changes the hydrophobic property of Tyr to hydrophilic, and the carboxylation of the amino group of Lys introduces a bulky hydrophilic group which will screen the hydrophobic cluster around Lys, similar to the guanidination of Lys described above. The modified enzymes were strongly thermostabilized.

5.3.2 Basis of thermostability in guanidinated proteins

Guanidination of pig heart lactate dehydrogenase produced the thermostabilized protein derivative (Minotani et al., 1979). One possible explanation for the enhanced stability is the decrease of thermodynamically unfavorable hydrophobicity of the protein surface. To search in more detail for a mechanism for protein stabilization, the following experiments were performed.

5.3.2.1 *Subunit–subunit interaction*

There exist two types of pig lactate dehydrogenase from heart and muscle, and both of them consist of four identical subunits, H and M, respectively (Holbrook et al., 1975). On treating the heart-type enzyme (H_4) with *o*-methylisourea, protein thermostability linearly increased with increasing number of modified Lys residues, up to an average of 5 mol Lys per subunit. On further modification of Lys the stability

Table 5.1—Thermostabilities of various isozymes from unmodified muscle-type and guanidinated heart-type lactate dehydrogenase. The values of t_m were estimated from the curves representing the residual activities after 30 minutes of heating at various temperatures. H and M denote the unmodified heart- and muscle-type enzymes, respectively. Thermal inactivation of all the enzyme samples was irreversible (Redrawn from Shibuya et al., 1982)

Average number of guanidinated lysine residues in H monomer	$t_m(°C)$				
	H_4	H_3M	H_2M_2	HM_3	M_4
0	61	56	54	55	58
4.7	65	62	59	57	—
15.1	70	66	63	60	—

increased only slowly, and 15 out of a total reactive 16 Lys per subunit were modified (Minotani et al., 1979). Such a gradual increase in thermostability with modification may not be due to a newly formed H-bond or salt bridge, but to increased surface hydrophilicity (Mozhaev et al., 1988a).

Various types of hybrid enzymes were prepared from the guanidinated H_4 with an average five or 15 modified Lys per subunit and native M_4, and the thermostabilities of the hybrids were compared (Shibuya et al., 1982). Thermostabilities of various hybrid enzymes are shown in Table 5.1.

Thermal inactivation of the enzyme samples obeyed first-order kinetics at any temperatures examined, and the Arrhenius plots for the first-order rate constant (k_r) were all linear. According to the transition-state theory (Eyring, 1935), the standard activation enthalpy (ΔH^*) was calculated from the equation $\Delta H^* = E_a - RT$, where R is the gas constant and E_a is the energy of activation estimated from the slope of empirical Arrhenius plot of k_r. The standrad activation free energy (ΔG^*) and the standard activation entropy (ΔS^*) were calculated from the equation $\Delta G^* = RT(\ln(\kappa T/h) - \ln k_r)$ and $\Delta G^* = \Delta H^* - T\Delta S^*$, respectively, where κ is Boltzmann's constant and h is Planck's constant (Eyring, 1935). Thermodynamic parameters, ΔG^*, ΔH^* and ΔS^* for the irreversible thermal inactivation of some hybrid enzymes are shown in Table 5.2.

From the results shown in the table, the following conclusions were deduced. The increased thermostability of H_4 with five guanidinated Lys per subunit ($^{G5}H_4$) is due to the decrease in ΔS^*, and more increased stability of $^{G14}H_4$ is due to the balance between increased ΔH^* (stabilization) and increased ΔS^* (destabilization).

Table 5.2—Thermodynamic parameters for thermal inactivation reactions of some isozymes of pig lactate dehydrogenase. The parameters were estimated from the Arrhenius plots of the thermal inactivation constant. $\Delta G^*, \Delta H^*$ and ΔS^* represent the standard activation energy, standard activation enthalpy and standard activation entropy, respectively. The values for H_4 and $^{G15}H_4$ were determined at 65°C and those for H_2M_2 and $^{G15}H_2M_2$ at 57°C. G5 and G15 denote enzyme samples with an average of 4.7 and 15.1 modified Lys per subunit, respectively (Redrawn from Shibuya et al., 1982)

Isozyme	ΔG^* (kcal mol^{-1})	ΔH^* (kcal mol^{-1})	ΔS^* (cal mol^{-1} deg^{-1})
H_4	23.7	97	217
$^{G5}H_4$	25.0	97	213
$^{G15}H_4$	26.1	114	260
H_2M_2	23.9	69	137
$^{G5}H_2M_2$	25.3	97	217

The stability of the Q-axis dimer (see Chapter 2) is governed by the number of salt bridges formed across the Q-axis, and the rate-determining step of thermal denaturation of lactate dehydrogenase comprises the distortion of dissociation of one of the two Q-contacts of the tetramer (Muller and Klein, 1981). Subunit interactions were also shown to make a contribution to urea sensitivity of the isozymes of bovine lactate dehydrogenase (Chan and Shanks, 1977), and the interactions between homologous subunits were considered to exhibit a greater stabilizing effect on sensitivity than the heterologous interactions.

As shown in Table 5.2, the difference in ΔG^* between $^{G5}H_4$ and H_4 at 65°C was almost the same as that between $^{G5}H_2M_2$ and H_2M_2 at 57°C. The ΔG^* difference between $^{G5}H_4$ and H_4 at 65°C was almost the same as that at 57°C. These results indicate that the thermostability of H_4 and H_2M_2 increases to almost the same extent on guanidination of the H monomer. This supports the suggestion that subunit–subunit interactions make a contribution to the thermostability of pig lactate dehydrogenase, and that the interactions between heterologous (and probably also homologous) subunits are affected by guanidination. The behavior of amino acid residues in the subunit–subunit contact area in lactate dehydrogenase is strikingly different for the three different types of contact generated by the molecular P-, Q- and R-axes (Holbrook *et al.*, 1975). Of the residues involved in forming the P-, Q- and R-axes contact surfaces, those in the P- and R-axes contact regions are changed much more than the Q-axis contact (Eventoff *et al.*, 1977). This may indicate that the contribution of heterologous interactions across the P- and R-axes to the thermostability of the hybrid isozymes is smaller than that across the Q-axis. On guanidination of H monomer, therefore, the contribution of heterologous interactions to the thermostability of the hybrid isozymes may be greatly changed with the contact surfaces across the P- and R-axes.

5.3.2.2 Compact packing

Thermophilic enzymes are considered to acquire thermostability while sacrificing the conformational flexibility essential for enzyme activity (Zuber, 1981). The flexibility (fluctuation or conformational motility) of proteins can be measured by the hydrogen–deuterium (H–D) exchange reaction rates of H-atoms of peptides and of Trp and Tyr residues in proteins (Gurd and Rothgeb, 1979). The technique has been shown to provide useful information on dynamic protein structure (Hvidt and Nielson, 1966; Englander, 1975; Nakanishi and Tsuboi, 1976, 1978; Nakanishi *et al.*, 1973, 1978; Ohta *et al.*, 1978, 1981; Yamada *et al.*, 1981; Gregory and Rosenberg, 1986). The conformational change of lactate dehydrogenase induced by guanidination was then examined by the H–D exchange reactions (Abe *et al.*, 1983).

Table 5.3 shows the changes of H–D exchange reaction of peptides and Tyr and Trp residues in pig heart lactate dehydrogenase on guanidination. In the table half-time represents half of the time necessary for complete H–D exchange, and fluctuation degree represents the ratio of H-atoms being accessible to solvent to the total H-atoms in peptides and Trp and Tyr residues. The table indicates that the Trp residues are freely accessible to solvent and that the fluctuation degrees (solvent accessibility) of Trp and Tyr residues and peptide bonds in the enzyme decrease in the order Trp > Tyr ≫ peptide. On guanidination of Lys in native enzyme, the

Table 5.3—Changes in half-life of H–D exchange reaction of peptides and tyrosine and tryptophan residues in pig heart lactate dehydrogenase on guanidination. The H–D exchange reaction of peptide was followed by the change of amide II absorbance (at $1550\,cm^{-1}$)/amide I absorbance (at $1750\,cm^{-1}$). The H–D exchange reaction of Tyr or Trp residues was followed by a change in ultraviolet absorption at 292.5 nm or a change in fluorescence intensity above 300 nm, respectively. The modified enzymes with an average of 4.7 or 15.1 guanidinated Lys per subunit are designated Enzyme 1 and 2, respectively. Fluctuation degree means the solvent accessibility of H atoms in peptides and Tyr and Trp residues. The results in the table indicate that Trp residues are freely accessible to solvent, but peptides are buried in the protein interior, so that they show an extremely low frequency of access to solvent (Redrawn from Abe et al., 1983)

	Half-life (ms)		
	Peptides (46°C)	Tyrosine residues (27°C)	Trytophan residues (27°C)
Native enzyme	2.4×10^7	38	30
Enzyme 1	6.0×10^7	41	26
Enzyme 2	2.6×10^8	60	30
Fluctuation degree	10^{-7}–10^{-8}	10^{-2}	1

fluctuation degree of Trp residues is not affected, suggesting that the state of the protein surface reflecting conformational flexibility is not changed. In the first step of thermostabilization (H_4 to $^{G5}H_4$) (Shibuya et al., 1982), only the conformation exhibiting the smallest fluctuation degree (peptide) is affected and becomes rigid. In the second step ($^{G5}H_4$ to $^{G15}H_4$), not only the peptides but also Tyr residues become less fluctuating. It can be said that protein stability accompanies the increased rigidity of the conformation of the protein interior, which is confirmed by the recent molecular reason for protein stability (see Chapter 3). The X-ray crystal structure of pig heart lactate dehydrogenase is known (Holbrook et al., 1975). If the guanidinated Lys residues are specified by the amino acid sequence analysis (or X-ray crystal analysis), the molecular reason for enhanced stability on guanidination of the enzyme could be deduced. Such attempts, however, were unsuccessful, because of non-specific modification of Lys residues in number and in position, as found the sequence analysis of $^{G5}H_4$.

REFERENCES

Abe, M., Nosoh, Y., Nakanishi, M. and Tsuboi, M. (1983) Hydrogen–Deuterium exchange studies on guanidinated pig heart lactate dehydrogenase. *Biochim. Biophys. Acta*, **746**, 176–181.

Alberts, B., Bray, D., Lewis, J., Raff, M., Roberts, K. and Watson, J. D. (1983) *Molecular biology of the cell.* Garland, New York.

Arakawa, T. and Timasheff, S. N. (1982) Stabilization of protein structure by sugars. *Biochemistry*, **21**, 6536–6544.

Branner-Yorgensen, S. (1983) Practical approaches to chemical modification of enzymes for industrial use. *Biochem. Soc. Trans.*, **11**, 20–21.

Bungay, H. R. and Belfort, G. (eds) (1987) *Advanced biochemical engineering.* Wiley, New York.

Burley, S. K. and Petsko, G. A. (1985) Aromatic–aromatic interaction: a mechanism of protein structure stabilization. *Science*, **229**, 23–28.

Chan, W. W.-C. and Shanks, K. E. (1977) Contribution of subunit interactions to the stability of lactate dehydrogenase. *J. Biol. Chem.*, **252**, 6163–6168.

Chibata, I., Tosa, T. and Sato, T. (1979) Use of immobilized cell systems to prepare fine chemicals. In *Microbial technology* (H. J. Peppler and D. Perlman, eds), pp. 433–461. Academic Press, London.

Chothia, C. (1984) Principles that determine the structure of proteins. *Annu. Rev. Biochem.*, **53**, 537–572.

Cohen, L. A. (1970) Chemical modification as a probe of structure and function. *In The enzymes* (P. D. Boyer, ed.), Vol. 1, pp. 147–211. Academic Press, New York.

Cupo, P., El-Deiry, W., Whitney, P. L. and Awad, W. H., Jr (1980) Stabilization of proteins by guanidination. *J. Biol. Chem.*, **255**, 10828–10833.

De Renobales, M. and Welch, W. Jr (1980) Chemical cross-linking stabilizes the enzymatic activity and quaternary structure of formyltetrahydrofolate synthetase. *J. Biol. Chem.*, **255**, 10460–10463.

Englander, S. M. (1975) Measurement of structural and free energy changes in hemoglobin by hydrogen exchange methods. *Ann. NY Acad. Sci.*, **244**, 10–27.

Eventoff, W., Rossmann, M. G., Taylor, S. S., Torff, H. -J., Meyer, H., Keil, W. and Klitz, H. -H. (1977) Structural adaptations of lactate dehydrogenase isoenzymes. *Proc. Natl. Acad. Sci. USA*, **74**, 2677–2681.

Eyring, H. (1935) Activated complex in chemical reactions. *J. Chem. Phys.*, **3**, 107–115.

Fojo, A. Y., Whitney, P. L. and Awad, W. H., Jr (1982) Effects of acetylation and guanidination of alkaline conformations of chymotrypsin. *Arch. Biochem. Biophys.*, **224**, 636–642.

Gregory, R. B. and Rosenberg, A. (1986) Protein conformational dynamics measured by hydrogen isotope exchange techniques. *Methods Enzymol.* (C. H. W. Hirs and S. N. Timasheff, eds), Vol. 131, pp. 448–508. Academic Press, New York.

Gurd, F. R. N. and Rothgeb, T. M. (1979) Motions in proteins. *Adv. Protein Chem.*, **33**, 73–165.

Hansch, C. and Leo, A. (1979) *Substituent constants for correlational analysis in chemistry and biology.* Wiley, New York.

Hirs, C. H. W. and Timasheff, S. N. (eds) (1977) Chemical modification. *Methods Enzymol.*, **47**, 407–498.

Holbrook, J. J., Liljas, A., Steindel, S. J. and Rossmann, M. G. (1975) Lactate dehydrogenase. *The enzymes* (P. D. Boyer, ed.), Vol. 11, pp. 191–292. Academic Press, New York.

Holmes, D. S. (1987) Molecular enzyme engineering. In *Advanced biochemical engineering* (H. R. Bungay and G. Belford, eds), pp. 129–165. Wiley, New York.

Hvidt, A. and Nielson, S. O. (1966) Hydrogen exchange in proteins. *Adv. Protein Chem.*, **21**, 287–386.

Jensen, S. E., Westlake, D. W. S. and Wolfe, S. (1984) Production of penicillins and cephalosporins in an immobilized enzyme reactor. *Appl. Microbiol. Biotech.*, **20**, 155–160.

Kagawa, Y. and Nukiwa, N. (1981) Conversion of stable ATPase to labile ATPase by acetylation and the $\alpha\beta$ and $\alpha\gamma$ subunit complexes during its reconstitution. *Biochem. Biophys. Res. Commun.*, **100**, 1370–1376.

Kaiser, E. T., Lawrence, D. S. and Rokita, S. E. (1985) The chemical modification of enzymatic specificity. *Annu. Rev. Biochem.*, **54**, 565–595.

Katchalski-Katzir, E. and Freeman, A. (1982) Enzyme engineering reaching maturity. *Trends Biochem. Sci.*, **7**, 427–431.

Katchalski-Katzir, E., Silman, I. and Goldman, R. (1971) Effect of the microenvironment on the mode of action of immobilized enzymes. *Adv. Enzymol.*, **34**, 445–536.

Klein, M. D. and Langer, R. (1986) Immobilized enzymes in clinical medicines: an emerging approach to new drug therapies. *Trends Biotech.*, **4**, 179–186.

Klibanov, A. M. (1979) Enzyme stabilization by immobilization. *Anal. Biochem.*, **93**, 1–25.

Klibanov, A. M. (1983) Immobilized enzymes and cells as practical catalyst. *Science*, **219**, 722–727.

Kricka, L. J. and Thorpe, G. H. G. (1986) Immobilized enzymes in analysis. *Trends Biotech.*, **4**, 253–258.

Krigbaum, W. R. and Komoriya, A. (1979) Local interactions as a structural determinant for protein molecules. II. *Biochim. Biophys. Acta*, **576**, 204–208.

Kristjansson, J. K. (1989) Thermophilic organisms as sources of thermostable enzymes. *Trends Biotech.*, **7**, 349–353.

Lundblad, R. L. and Noyes, C. M. (1985) *Chemical reagents for protein modification.* CRC Press, Boca Raton.

Marshall, J. J. (1978) Manipulation of the properties of enzymes by covalent attachment of carbohydrate. *Trends Biochem. Sci.*, **3**, 79–83.

Martinek, K. and Mozhaev, V. V. (1985) Immobilization of enzymes: an approach to fundamental studies in biochemistry. *Adv. Enzymol.*, **57**, 179–249.

Means, G. E. and Feeney, R. E. (1971) *Chemical modification of proteins.* Holden-Day, San Francisco.

Minotani, N., Sekiguchi, T., Bautista, J. G. and Nosoh, Y. (1979) Basis of thermostability in pig heart lactate dehydrogenase treated with *o*-methylisourea. *Biochim. Biophys. Acta*, **581**, 334–341.

Moore, B. R. and Free, S. J. (1985) Protein modification and its biological role. *Int. J. Biochem.*, **17**, 283–289.

Mosbach, K. (ed.) (1976) Immobilized enzymes. *Methods Enzymol.*, **44**, 3–936.

Mosbach, K. (ed.) (1987) Immobilized enzymes and cells. *Methods Enzymol.*, **135**, 3–604.

Mounter, L. A., Shipley, B. A. and Mounter, M. E. (1963) The inhibition of hydrolytic enzymes by organophosphorous compounds. *J. Biol. Chem.*, **238**, 1979–1983.

Mozhaev, V. V. and Martinek, K. (1984) Structure–stability relationship in proteins: new approach to stabilizing enzymes. *Enzyme Microb. Technol.*, **6**, 50–59.

Mozhaev, V. V., Berezin, H. V. and Martinek, K. (1988a) Structure-stability relationship in proteins: fundamental tasks and strategy for the development of stabilized enzyme catalysis for biotechnology. *CRC Crit. Rev. Biochem.*, **23**, 235–281.

Mozhaev, V. V., Siksnis, V. A., Melik-Nubarov, N. S., Galkantaite, N. Z., Denis, G. J., Butkus, E. P., Zaslavsky, B. Yu, Mestechkina, N. M. and Martinek, K. (1988b) Protein stabilization via hydrophilization: covalent modification of trypsin and α-chymotrypsin. *Eur. J. Biochem.*, **173**, 147–154.

Muller, J. (1981) Stability of lactate dehydrogenase. I. Chemical modification of lysines. *Biochim. Biophys. Acta*, **669**, 210–221.

Muller, J. and Klein, C. (1981) Stability of lactate dehydrogenase. II. Hybrids and geometric isomers. *Biochim. Biophys. Acta*, **671**, 38–43.

Nakanishi, M. and Tsuboi, M. (1976) Structure and fluctuation of a streptomyces subtilisin inhibitor. *Biochim. Biophys. Acta*, **434**, 365–376.

Nakanishi, M. and Tsuboi, M. (1978) Measurement of hydrogen exchange at the tyrosyl residues in ribonuclease A by stopped-flow and ultraviolet spectroscopy. *J. Am. Chem. Soc.*, **100**, 1273–1275.

Nakanishi, M., Tsuboi, M. and Ikegami, A. (1973) Fluctuation of the lysozyme structure. II. Effect of temperature and binding of inhibitors. *J. Mol. Biol.*, **75**, 673–682.

Nakanishi, M., Nakamura, H. and Tsuboi, M. (1978) Hydrogen–deuterium exchange rate between a peptide group and an aqueous solvent as determined by a stopped-flow ultraviolet spectrophotometery. *Bull. Chem. Soc. Japan*, **51**, 1988–1990.

Nieto, M. A. and Palacian, E. (1983) Effects of temperature and pH on the regeneration of the amino groups of ovalbumin after modification with citraconic and dimethyl-maleic anhydride. *Biochim. Biophys. Acta*, **749**, 204–210.

Nosoh, Y. and Sekiguchi, T. (1988) Protein thermostability: mechanism and control through protein engineering. *Biocatalysis*, **1**, 257–273.

Ohta, S., Nakanishi, M., Tsuboi, M., Yoshida, M. and Kawaga, Y. (1978) Kinetics of hydrogen–deuterium exchange in ATPase from a thermophilic bacterium PS3. *Biochem. Biophys. Res. Commun.*, **80**, 929–935.

Ohta, S., Tsuboi, M., Yoshida, M. and Kagawa, A. (1981) Intersubunit interactions in proton translocating adenosine triphosphatase as revealed by hydrogen-exchange kinetics. *Biochemistry*, **19**, 2160–2165.

Peters, K. and Richards, F. M. (1979) Chemical cross-linking: reagents and problems in studies of membrane structure. *Annu. Rev. Biochem.*, **46**, 523–551.

Privalov, P. L. (1979) Stability of proteins. *Adv. Protein Chem.*, **33**, 167–241.

Schmid, R. D. (1979) Stabilized soluble enzymes. *Adv. Biochem. Eng.*, **12**, 41–118.

Sekiguchi, T., Oshiro, S., Goingo, E. M. and Nosoh, Y. (1979a) Chemical modification of ε-amino groups in glutamine synthetase from *Bacillus stearothemophilus* with ethyl acetimidate. *J. Biochem. (Tokyo)*, **85**, 75–78.

Sekiguchi, T., Oshiro, S., Goingo, E. M. and Nosoh, Y. (1979b) Cross-linking with diimidates of glutamine synthetase from *Bacillus stearothermophilus*. *J. Biochem. (Tokyo)*, **86**, 447–452.

Shami, E. Y., Rothstein, A. and Ramjeesingh, M. (1989) Stabilization of biologically active proteins. *Trends Biotech.*, **7**, 186–190.

Shatsky, M. A., Ho, H. C. and Wang, J. H. (1973) Stabilization of glycogen phosphorylase b by reductive alkylation with aliphatic aldehyde. *Biochim. Biophys. Acta*, **303**, 298–307.

Shaw, E. (1970) Chemical modification by active-site-directed reagents. In *The enzymes* (P. D. Boyer, ed.), Vol. 1, pp. 91–146. Academic Press, New York.

Shibuya, H., Abe, M., Sekiguchi, T. and Nosoh, Y. (1982) Effect of guanidination on subunit interactions in hybrid isozymes from pig lactate dehydrogenase. *Biochim. Biophys. Acta*, **708**, 300–304.

Torchilin, V. P. and Martinek, K. (1979) Enzyme stabilization without carriers. *Enzyme Microb. Technol.*, **1**, 74–82.

Torchilin, V. P., Maksimenko, A. V., Smironov, V. N., Berezin, I. V., Klibanov, A. M. and Martinek, K. (1978) The principles of enzyme stabilization. III. The effect of the length of intramolecular cross-linkages on thermostability of enzymes. *Biochim. Biophys. Acta*, **522**, 277–283.

Torchilin, V. P., Maksimenko, A. V., Smirnov, V. N., Berezin, I. V. and Martinek, K. (1979) The principles of enzyme stabilization. IV. 'Key' functional groups in tertiary structure of proteins and the effect of their modification on thermostability of enzymes. *Biochim. Biophys. Acta*, **567**, 1–11.

Trevan, M. D. (1980) *Immobilized enzymes*. Wiley, London.

Tsai, C. S., Tsai, Y. H., Lauzon, G. and Cheng, S. T. (1974) Structure and activity of methylated horse liver alcohol dehydrogenase. *Biochemistry*, **13**, 440–443.

Tuengler, P. and Pfleiderer, G. (1977) Enhanced heat, alkaline and tryptic stability of acetamidinated pig heat lactate dehydrogenase. *Biochim. Biophys. Acta*, **484**, 1–8.

Ulmer, K. M. (1983) Protein engineering. *Science*, **219**, 666–676.

Urabe, I., Nanjo, H. and Okada, H. (1973) Effect of acetylation of *Bacillus subtilis* α-amylase on the kinetics of heat inactivation. *Biochim. Biophys. Acta*, **302**, 73–79.

Urabe, I., Yamamota, M., Yamada, Y. and Okada, H. (1978) Effect of hydrophobicity of acyl groups on the activity and stability of acylated thermolysin. *Biochim. Biophys. Acta*, **524**, 435–441.

Vandamme, E. J. (1983) Peptide antibiotic production through immobilized biocatalyst technology. *Enzyme Microb. Technol.*, **5**, 403–415.

Wolfenden, R., Anderson, L., Cullis, P. and Southgate, C. (1981) Afinities of amino acid side chains for solvent water. *Biochemistry*, **20**, 849–855.

Wood, L. L. and Calton, G. J. (1984) A novel method of immobilization and its use in aspartic acid production. *Bio/technology*, **2**, 1080–1084.

Yamada, T., Shimizu, H., Nakanishi, M. and Tsuboi, M. (1981) Environment of the tryptophan residues in a myosin head: a hydrogen–deuterium exchange study. *Biochemistry*, **20**, 1162–1168.

Zaborsky, O. R. (1973) *Immobilized enzymes*. CRC Press, Boca Raton.

Zuber, H. (1981) Structure and function of thermophilic enzymes. In *Structural and functional aspects of enzyme catalysis* (H. Eggerer and R. Huber, eds), pp. 114–127. Springer-Verlag, Berlin.

6

Stabilization through biological modification

This strategy for stabilizing proteins is based on the *in vivo* and *in vitro* substitution of nucleotide(s) of a gene encoding a protein, which produces a mutant protein with substituted amino acid(s) (see section 2.2.1). Therefore, the principle of this strategy is entirely different from those of chemical and physical modification techniques for protein stabilization as described above (Chapter 5). Most of the target proteins for biological modification have been the enzymes of which X-ray crystal structures are known, and the tertiary structures of their modified proteins can also be rather easily determined. Thus, the difference in stability between the parent and modified enzymes can be discussed in terms of the different conformational changes between the two proteins. In other words, biological modification might also be a useful tool for analyzing the mechanism of protein stability.

There are two kinds of procedures for the biological modification of proteins. One is classical mutagenesis effected by the application of chemicals or by irradiation with ultraviolet (UV) or X-rays to organisms (Miller, 1972). Such mutagenesis has long been used as a tool for analyzing gene function and for altering enzyme activity. There are, however, several drawbacks to classical mutagenesis. For example, because of random mutations in the genetic material, a mutation in the gene of interest occurs only by chance. In addition, it is extremely laborious or even impossible to isolate a desirable mutant strain. One point mutation, however, can be introduced into an isolated gene (or the plasmid harboring the gene) by various mutagens; e.g. with kanamycin nucleotidyltransferase (Matsumura and Aiba, 1985) and even into a gene in organisms; e.g. yeast iso-1-cytochrome c (Das *et al.*, 1989). Another modern procedure is site-directed mutagenesis, which became possible by recent advances in genetic engineering due to the remarkable development of recombinant DNA techniques (Hanna, 1987). Site-directed mutagenesis permits a single or a few selected amino acid residues in a specific enzyme to be precisely altered or replaced, eliminating the problems inherent in classical mutagenesis. The latter technique using *in vitro* mutagenesis is generally called protein engineering (Ulmer, 1983). Protein engineering is now becoming a promising tool for stabilizing proteins and also for revealing the mechanism of protein stability (see section 3.2).

6.1 SITE-DIRECTED MUTAGENESIS

Site-directed mutagenesis consists of the following principal steps: (1) isolation of the gene coding for a desired enzyme; (2) preparation of a single-stranded plasmid containing the enzyme gene (Dillon *et al.*, 1985; Hanna, 1987); (3) an internal mismatched oligonucleotide complementary to the region to be altered is hybridized to the single-stranded DNA; (4) a double-stranded heteroduplex plasmid is synthesized by DNA polymerase and ligase; and (5) the double-stranded DNA is transformed into *E. coli*, resulting in two classes of the parental and those carrying the site-directed mutation. Recently an improved method was proposed to provide a very strong selection against the non-mutagenized strand of a double-stranded DNA (Kunkel, 1985; Kunkel *et al.*, 1987). A single-stranded plasmid incorporating uracil in DNA is synthesized in a dut^- ung^- double mutant bacterium. This uracil-containing strand is used as a template for the synthesis of a double-stranded DNA. When the resulting double-stranded DNA is transformed into a cell with uracil N-glycosylase, the uracil-containing strand is inactivated with high efficiency. Thus, typical mutagenesis frequencies are greater than 50%. In Fig. 6.1 are shown the steps of the above procedure.

The important aspects of site-directed mutagenesis are the modification of various functions of proteins. Using site-directed mutagenesis, the following kinds of enzyme functions, for instance, could be introduced in a controlled and predictable fashion (Ulmer, 1983):

(1) kinetic properties, including the turnover number of the enzymes and the Michaelis constant for a particular substrate;
(2) thermostability and temparature optimum;
(3) stability and activity in non-aqueous solvents;
(4) substrate and reaction specificity;
(5) cofactor requirements;
(6) pH optimum;
(7) protease resistance;
(8) allosteric regulation;
(9) molecular weight and subunit structure;
(10) incorporation of unnatural amino acids.

Some of the enzyme functions have already been modified through protein engineering: e.g. the specific activity of tyrosyl-tRNA synthetase (Winter *et al.*, 1982; Wilkinson *et al.*, 1983, 1984; Fersht *et al.*, 1985; Wells and Fersht, 1985), of protease (Carter and Wells, 1988) and of alkaline phosphatase (Ghosh *et al.*, 1986), and quantitative changes in substrate specificity of trypsin (Craik *et al.*, 1985) and of subtilisin (Wells *et al.*, 1987). A qualitative change in substrate specificity was recently reported with lactate dehydrogenase; i.e. the specificity was converted from lactate to malate (Wilks *et al.*, 1988).

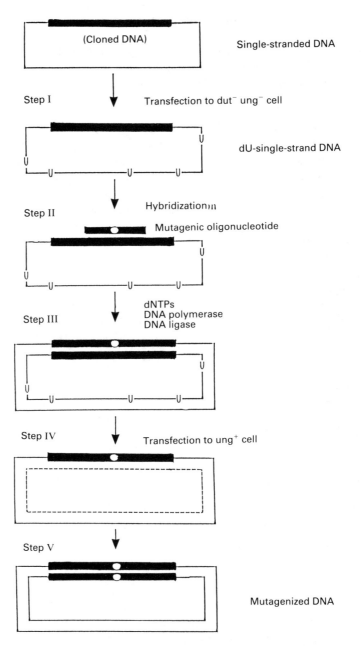

Fig. 6.1—Scheme for site-directed mutagenesis based on the method of Kunkel (Kunkel, 1985; Kunkel et al., 1987), which uses a DNA template containing a small number of uracil residues in place of T. The uracil-containing DNA is produced in E. coli $dut^-\ ung^-$ strain (step I). The broad line represents the gene coding for an enzyme to be engineered. The uracil-containing single-stranded DNA is hybridized with a mutagenic oligonucleotide as primer (step II). The annealed DNA is used *in vitro* for the synthesis of a complementary strand using dNTPs, DNA polymerase and DNA ligase (step III). When the double-stranded DNA is transfected into *E. coli* ung^+ with uracil-*N*-glycosylase, the uracil-containing strand is inactivated with high efficiency (step IV), and the majority of progeny arises from the complementary strand which contains the desired alteration (step V). U, incorporated uracil; ○, mutation site; ----, inactivated uracil-containing strand.

6.2 MUTAGENESIS WITH KNOWLEDGE OF PROTEIN STRUCTURE

It is now an easy task to substitute amino acid residues in a protein at will. To predict the effects of amino acid substitution on protein function (protein design), however, the information on the relationship between function and structure, especially X-ray crystal structure, of the protein are indispensable. The target enzymes described above are all those whose structure–function relationships are known.

6.2 MUTAGENESIS WITH KNOWLEDGE OF PROTEIN STRUCTURE

If a protein is designed to be stabilized through site-directed mutagenesis, the informations on structure–stability relationship must be at hand. As described in Chapters 3 and 4, we have not yet any generalized molecular reasons for protein stability, but several mechanisms are now available.

Recent advances in recombinant DNA technique permit the isolation and manipulation of a desired gene (Hanna, 1987). The isolated gene is the source for the rapid sequencing of a protein coded by the gene, because it is much easier to sequence the DNA of the gene than the protein itself. Thus, the amino acid sequence of a protein can be easily and correctly determined. The most requisite information on protein conformation for engineering a protein, however, is its X-ray crystal structure. But due to the difficulty in crystallizing proteins, such structural determinations are now limiting steps in the advance of protein engineering. In addition, the thremodynamic and kinetic analyses for protein denaturation will provide more detailed, quantitative information on the mode of stabilizing effect of substituted amino acid residues on protein stability (Chapter 3; see also section 6.2.3). Thermodynamic analysis, however, can be made with reversible denaturation of monomeric proteins, but not with irreversible denaturation (Chapter 3). Most proteins, especially oligomeric, denature irreversibly.

Thus, the following monomeric enzymes have been widely used for stabilizing and analyzing protein stability through protein engineering: T4 phage lysozyme and subtilisin BPN'. In this chapter, we describe in detail some typical examples of protein engineering for stabilizing proteins, including lysozyme and subtilisin.

6.2.1 Altering destabilizing factors

As described in Chapter 3, the free energy change (ΔG) for protein unfolding is expressed by the following equation: $\Delta G = \Delta H - T\Delta S$. Here, ΔH and ΔS are the enthalpy and entropy changes for unfolding. This equation indicates that the degree of conformational stability of protein (ΔG, negative) depends on the balance between the stabilizing (ΔH, negative) and destabilizing factors (ΔS, negative). Thus enhancing the stabilizing factors or reducing the destabilizing factors will stabilize proteins. The strategies for reducing the destabilizing factors is to form disulfide bond(s) in the protein molecule which will decrease the entropy of the unfolded state, and also to substitute amino acid residues to cause the entropy decrease of the unfolded state (Chapter 3). Another strategy for stabilizing proteins is to increase the stabilizing factors such as hydrophobic and electrostatic interactions and hydrogen bonding. The stabilization of enzymes by reducing the destabilizing factors is first described.

6.2.1.1 Introduction of disulfide bonds

In globular proteins, disulfide bonds provide conformational stability (Thronton, 1981; Richardson, 1981; see also Chapter 3). Theoretical treatments predicted that a cross-link such as a disulfide bond should stabilize proteins, depending on the length of primary chain bridged by the bond, owing to a decrease in the chain entropy of the unfolded state of the cross-linked form (Poland and Scheraga, 1965a). The increase in chain entropy is the predominant destabilizing factor to the free energy of stabilization, so that cross-linking is expected to shift the unfolding equilibrium toward the native state. These calculations are supported by thermodynamic studies on cross-linked proteins (e.g. Johnson *et al.*, 1978; Goto and Hamaguchi, 1979; Lin *et al.*, 1984; Ueda *et al.*, 1985).

There are relatively few naturally occurring proteins containing a single disulfide bond. The presence of more than one disulfide bond complicates quantification of the stabilizing factors of each bond. Protein engineering has made it possible to construct proteins containing a single disulfide bond, which facilitates the characterization of the contributions of the cross-link to protein stability. The contribution of the disulfide bond to protein stability can be thermodynamically estimated only with reversible unfolding reactions. Most naturally occurring proteins, however, denature irreversibly, and in such cases protein stability is defined by resistance to irreversible processes. Most proposed mechanisms for irreversible denaturation, *in vivo* and *in vitro*, are that unfolded proteins, before refolding, become susceptible to inactivation by damage due to heat or chemical or protease attack (Klibanov, 1983). Thus, attempts have been made to relate protein stability to irreversible denaturation or to relate its stability with reversible unfolding, as well as to determine the role of the disulfide bond in both processes (see the following reviews: Wetzel, 1987; Matthews, 1987; Shortle, 1989).

T4 phage lysozyme

The effect of engineered disulfide bonds on protein stability was first examined by Perry and Wetzel (1984). The three-dimensional structure of T4 phage lysozyme is well known (Fig. 6.2) and it has two unpaired Cys residues at positions 54 and 97 (Matthews and Remington, 1974; Rossmann and Argos, 1976; Remington *et al.*, 1978; Matthews *et al.*, 1981). The degree of conformational stability of cross-linked proteins was suggested to increase with the size of cross-linked loop (Thronton, 1981). Perry and Wetzel (1984) investigated the crystal structure of T4 lysozyme for sets of amino acids widely separated in primary sequence, but with side chain β-carbons lying in close proximity (within 5 Å). Simulated disulfides identical in structure to the natural protein disulfides (Thronton, 1981; Richardson, 1981) were examined by computer modeling of the enzyme. Several candidates were considered to be similar to the natural disulfides, and one of them was the simulated disulfide bond between the residues at position 3 and 97. This simulated bond has the following values for its structural parameters: χ_1, 117°; χ_2, 25°, χ_3, 134°; $\chi_{2'}$, 43°; $\chi_{1'}$, $-162°$; C_α–C_α distance, 5.71 Å; and C_β–C_β distance, 4.49 Å. Since residue 97 in the lysozyme is Cys, the replacement of the amino acid at residue 3 (Ile) with Cys may yield a new disulfide bond between positions 3 and 97 (Fig. 6.3).

Ile at position 3 in the enzyme was replaced with Cys by site-directed mutagenesis,

Ch. 6] **6.2 Mutagenesis with Knowledge of Protein Structure** 145

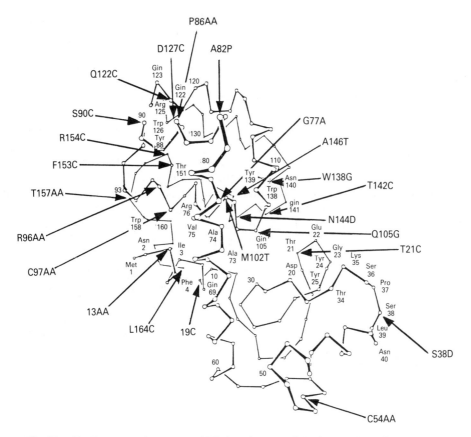

Fig. 6.2—The X-ray crystal structure of T4 phase lysozyme (Matthews and Remington, 1974). Some of the amino acid substitutions that have been made for analyzing protein stability and stabilizing proteins are indicated by arrows, with substituting positions. These substitutions are shown in the text.

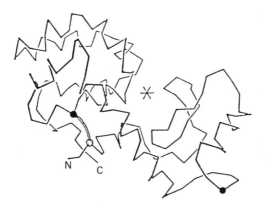

Fig. 6.3—The α-carbon backbone structure of T4 lysozyme. N and C represent the amino- and carboxy-termini, respectively, and the star the active site. Two natural Cys 54 and Cys 97 (●) and engineered Cys 3 (○) are shown. Cys 3 and Cys 97 are connected by a schematic disulfide (See also Fig. 6.6). (Modified from Perry and Wetzel, 1984.)

using the synthetic oligonucleotide pAGAATTATGAATTGTTTTGAAATGTTA (TGT is a codon for Cys, and a codon for Ile is TAT) (Perry and Wetzel, 1984). The gene coding for the mutant enzyme was expressed in *E. coli*.

The mutant protein with three free Cys residues exhibited the same specific activity and stability against irreversible thermal inactivation as found in the wild-type enzyme. Mild oxidation of the mutant protein with sodium tetrathionate or glutathione redox buffer generated the disulfide bond between Cys 3 (from Ile) and Cys 97. The oxidized mutant thus obtained also exhibited the same activity as those of the wild and mutant (reduced) enzymes. The oxidized mutant protein, however, showed enhanced thermostability (Fig. 6.4). The initial decay in activity in the wild-type had a half-life of 11 min, while the disulfide bonded protein had a half-life of 28 min. More dramatically, the activity of the disulfide bonded enzyme did not fall below 50% of the starting activity. In contrast, the wild-type enzyme exhibited only about 0.2% of its starting activity after 180 min. It was suggested that the additional stability of the disulfide form of enzyme is due to the cross-linkage and not to the amino acid substitution alone.

The thermostability of the mutant lysozyme with the engineered disulfide bond was further enhanced by chemical blocking or by mutagenic replacement of unpaired Cys at position 54 to Val or Thr in the disulfide bonded enzyme (Perry and Wetzel, 1986). One mode of decay of the wild-type enzyme was shown to be the oxidation of its two unpaired Cys residues at positions 54 and 97 (Perry and Wetzel, 1987).

Although the cross-link between two Cys residues at positions 3 and 97 provides free energy of stabilization of T4 lysozyme, detailed thermodynamic and kinetic

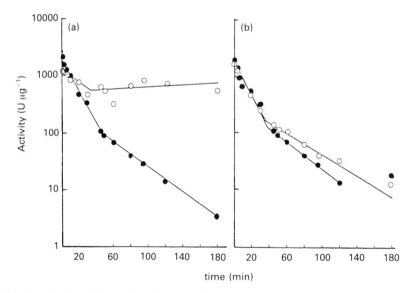

Fig. 6.4—Kinetics of thermal inactivation of T4 lysozyme derivatives at 67°C. (a) Wild-type enzyme (●) and mutant enzyme with a disulfide bond between Cys 3 and Cys 97 (○); (b) wild-type enzyme (●) and mutant enzyme with free Cys 3 and Cys 97 (○). (Redrawn from Perry and Wetzel, 1984.)

6.2 Mutagenesis with Knowledge of Protein Structure

analyses for thermal inactivation will reveal the mode of stabilization effect of the disulfide on the enzyme (Wetzel, 1987; Wetzel et al., 1988).

When the thermal inactivation conditions are carefully chosen, the thermodynamic stability (stability to reversible inactivation) of the wild-type and various mutant types of T4 lysozyme can be determined quantitatively (Backtel and Baase, 1987; Wetzel et al., 1988). In Table 6.1 are shown some thremodynamic properties of a series of T4 lysozymes. In general, introduction of a disulfide bond between two Cys residues at positions 3 and 97 stabilizes the enzyme against reversible unfolding (thermodynamic stabilization), although to a slight extent. The irreversible thermal inactivation of the enzyme is more stabilized on forming the disulfide bridge (Wetzel et al., 1988) (Fig. 6.5).

The irreversible denaturation of T4 lysozyme with and without the cross-link appears to be different (Wetzel et al., 1988). Since free Cys residues in T4 lysozyme are air oxidized on heating the enzyme (Perry and Wetzel, 1987), the two Cys residues at positions 54 and 97 are replaced by Val and Ser, respectively. This double mutant irreversibly lost its activity on exposure to 70°(°C), but the lost activity was restored almost completely by the addition of guanidine hydorchloride. This may be indicative of aggregation or mis-unfolding of irreversibly inactivated product(s). The cross-linked enzyme with Cys to Val replacement at position 54 was also thermally inactivated. The lost activity, however, did not exhibit any guanidine hydrochloride recovery. The cross-linked mutant may have received some chemical damage (Zale and Klibanov, 1986). It is thus possible that the disulfide introduced in T4 lysozyme may stabilize the enzyme against irreversible inactivation by restricting the unfolded state

Table 6.1 — The values of T_m of T_4 lysozyme mutants with disulfide bond at various positions. The difference between free energy for unfolding of the mutant and that of wild-type at 70°C from the thermodynamic analysis of the reversible melting temperature curves. Melting profiles were obtained by CD spectral change (Redrawn from Wetzel et al., 1988)

Mutant	T_m(°C)	$\Delta\Delta G$ (kcal mol^{-1})
Wild-type	65	0
V54(←C)	63	−0.7
T54(←C)	66	0.3
H96(←R)	58	−2.8
T146(←A)	59	−1.5
V54(←C), S97(←C)	59	−2.1
C3(←I)–C97	68	1.2
C3(←I)–C97, V54(←C)	66	0.4
C3(←I)–C97, T54(←C)	69	1.5
C3(←I)–C97, T54(←C), H96(←R)	60	−2.5
C3(←I)–C97, T54(←C), T146(←A)	62	−0.5

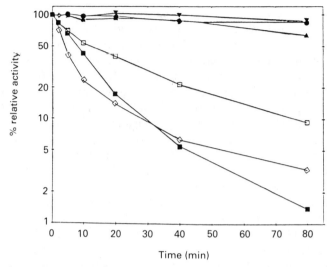

Fig. 6.5—Thermal inactivation of T4 lysozyme derivatives at 73°C. Mutant enzymes with Thr 54 (←Cys) (■), with a disulfide bond between Cys 3 (←Ile) and Cys 97 and with Thr 54 (←Cys) (●), with a disulfide bond between Cys 3 (←Ile) and Cys 97, Thr 54 (←Cys) and His 96 (←Arg) (▲), and with a disulfide bond between Cys 3 (←Ile) and Cys 97, Thr 54 (←Cys) and Thr 146 (←Ala) (▼). The wild-type (◇) and mutant with Thr 54 (←Cys) (□) in the presence of dithiothreitol, which block oxidative inactivation (Perry and Wetzel, 1987). (Redrawn from Wetzel et al., 1988.)

to a class of more compact structures. The mechanism essentially utilizes the constraining action of the disulfide for its effect on the properties of the thermally unfolded state rather than for its effect on the thermodynamics of the unfolding equilibrium.

Inspection for possible disulfide bonding sites in T4 lysozyme showed a best candidate for the site between positions 3 and 97 (Perry and Wetzel, 1984). Recently, Matsumura et al., (1989) re-examined the sites by theoretical calculations and computer modeling, and selected the sites between positions 9 and 164, between positions 21 and 142, between positions 90 and 122, and between positions 127 and 154 (Fig. 6.6). The cross-linking between positions 3 and 97 was found to be one of the best candidates. The relevant conformational parameters for disulfide bonds are summarized in Table 6.2.

The engineered Cys residues formed disulfide bonds on exposure to air *in vitro*. All the oxidized lysozymes were more stable than the corresponding reduced, non-cross-linked proteins towards thermal denaturation. Relative to wild-type enzyme, the values of T_m of the 9–164 and 21–142 mutants were increased by 6.4°C and 11.0°C, but the other mutants were either less stable or equally stable. The less stable mutants tend to have less favorable cross-links in the native structure, as measured by the equilibrium constants for the reduction of the engineered disulfide bonds by dithiothreitol. In common, the two disulfide bridges that are most effective in increasing the stability of T4 lysozyme have a large loop size and a location that includes a flexible part of the molecule (Fig. 6.6). From the results the authors suggest that stabilization due to the cross-link reducing the entropy of the unfolded state can

6.2 Mutagenesis with Knowledge of Protein Structure

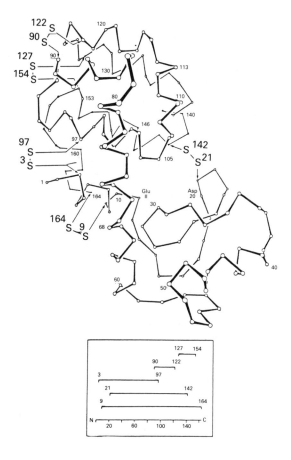

Fig. 6.6—Backbone structure of T4 lysozyme indicating locations of the engineered disulfide bonds. Four pairs of amino acid residues, Ile 9 and Leu 164, Thr 21 and Thr 142, Ser 90 and Gln 122, and Asp 127 and Arg 154, were all replaced with Cys, respectively. The mutant with a disulfide bond between Cys 3 (←Ile) and Cys 97 and with Thr 54 (←Cys) is also shown. The double introduction of Cys affected less or no activity, except for the mutants with Cys 21 and Cys 142 and with Cys 127 and Cys 154: about 49% and 20% reduction in activity with the former and the latter, respectively. The mutants with a disulfide bond between Cys 21 and 142, between Cys 90 and 122, and between Cys 127 and 154 decreased activity by 100%, 40% and 50%, respectively. Glu 11 and Asp 20 are the active site residues. The loop size for each disulfide bond is shown in the box. Conformational parameters for these engineered disulfide bonds are shown in Table 6.2. (Redrawn from Matsumura et al., 1989.)

be offset by the strain energy associated with formation of the disulfide bond in the folded state.

Refined crystal structure of T4 lysozyme cross-linked between the positions 9 and 164 was compared with that of the wild-type enzyme (Pjura et al., 1990). The wild-type and cross-linked enzymes have a very similar overall crystallographic temperature factor. This indicates that the introduction of the cross-link does not impose rigidity of the folded protein structure. In particular, the peptide chain from residues 162 to

Table 6.2 — Parameters for disulfide bonds in T_4 lysozyme. The candidate pairs of residues to be engineered were screened to find pairs that satisfy the following criteria. (1) The residues must be separated by at least 20 residues along the polypeptide chain, both residues must have γ-atoms, and these γ-atoms must be closer together than 6 Å. (2) The strain associated with formation of the disulfide bond must be minimized. The energy minimization, with defined parameters (Levitt, 1983) was confined to a small *molten zone* that included one residue on each side of the pair of the residues being replaced, allowing only the six residues to move. The strain energy was calculated as the difference between the energy of the protein with and without disulfide bond. The energy of the cross-linked form was found to be higher than that with uncrosslinked form by at least 4 kcal mol^{-1}. (3) Stabilizing interactions, such as hydrophobic interactions, H-bonds and van der Waals contacts that existed in the wild-type structure are analyzed. The introduction of Cys pairs should minimize the loss of pre-existing interactions that stabilize the wild-type structure. Final candidates were examined by simulating the disulfide bridges with the use of computer graphics. Geometry is the discrepancy with best-corresponding disulfide (Bernstein et al., 1977). Side-chain burial is the fraction of the substituted side chain in the wild-type structure inaccessible to solvent. Main chain B-value is the average crystallographic thermal factor of the N, C_α, C and O atoms of the substituted residue (Redrawn from Matsumura et al., 1989)

	I9/L164	T21/T142	S90/Q122	D127/R154	I3/C97
Geometry (Å)	1.35	1.33	0.94	1.10	0.93
Distance (Å)					
C^α–C^α	7.4	8.1	6.2	6.0	5.6
C/O^γ–C/O^γ	4.4	4.2	3.8	4.4	4.8
Strain energy (kcal mol^{-1})	4.1	20.9	6.7	8.4	7.4
Loop size (residues)	156	122	33	28	95
Loss of contacts	7	3	5	5	7
Side chain burial (%)	71, 69	19, 82	72, 34	0, 56	84, 78
Main chain B-value (Å)	17, 76	29, 14	14, 22	19, 21	15, 11

164 in the most liable part of the native molecule retains high mobility in the engineered protein, which is consistent with the idea that stabilization of the protein is due to the effect of the disulfide bond on the unfolded rather than the folded state.

The following precautions in stabilizing proteins by the engineered disulfide bond should be taken (Matsumura et al., 1989): (1) introduction of engineered disulfide should minimize the disruption or loss of interactions that stabilize the native structure; (2) the size of the loop formed by a disulfide bond should be as large as possible; and (3) the strain energy introduced by the engineered disulfide bond should be kept as low as possible.

Subtilisin BPN'

The high-resolution X-ray structure of subtilisin BPN' has been solved (Wright et al., 1969; Drenth et al., 1972; Bott et al., 1988), and the gene has been cloned and expressed in a secreted form in B. subtilis (Wells et al., 1983). Subtilisin contains no Cys residues. This simplifies the analyses of disulfide formation and will extend the generality of engineering the disulfide bond. Detailed studies with subtilisin were then attempted to examine the effects of engineered disulfide bonds on protein stability (Wells and Powers, 1986; Pantoliano et al., 1987; Mitchison and Wells, 1989).

Wells and Powers (1986) found the possible introduction sites of engineered disulfides in subtilisin by computer graphic analyses of the protein. Two pairs of model Cys residues can be built to come within good disulfide bond distance of each other (2.0 Å) (Richardson, 1981; Thronton, 1981), without making unfavorable contacts with neighboring atoms, or requiring main chain movement (Fig. 6.7). A model disulfide between positions 22 and 87 can be built with a reasonable left-handed geometry ($\chi_3 = -90°$), and another between positions 24 and 87 with right-handed geometry ($\chi_3 = +90°$). The conformations of the main chain involving the engineered disulfides are random coil (Fig. 6.8), and the introduced disulfides are 24 Å away from the catalytic site. In addition, this region has the advantage of being accessible so that the disulfides may react more readily with oxidizing or reducing agents. The surface accessibility should facilitate the analysis of the disulfide bond strength against reduction by, for example, dithiothreitol in the non-denatured protein. When the sets of amino acids at positions 22 (Thr) and 87 (Ser) or at positions 24 (Ser) and 87 are all replaced by Cys, the disulfide bond was formed *in vivo* between positions 22 and 87 or between positions 24 and 87, respectively. These engineered disulfide bonds have no detectable effects on the specific activities and the pH optima of the cross-linked enzymes.

Studies of autoproteolytic inactivation of wild-type subtilisin showed a relationship between autolytic and conformational stabilities of the protein (Wells and Powers, 1986). The autolysis of subtilisin on heating has been reported (Takahashi and Sturtevant, 1981). The cross-linked mutant between positions 24 and 87 and wild-type

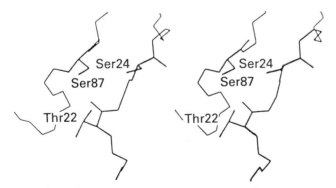

Fig. 6.7—Stereo view of the backbone structure of subtilisin showing residues Thr 22, Ser 24 and Ser 87 replaced with Cys to form disulfide bonds between positions 22 and 87 and between 24 and 87. (Redrawn from Wells and Powers, 1986.) Whole structure of subtilisin is shown in Fig. 6.8.

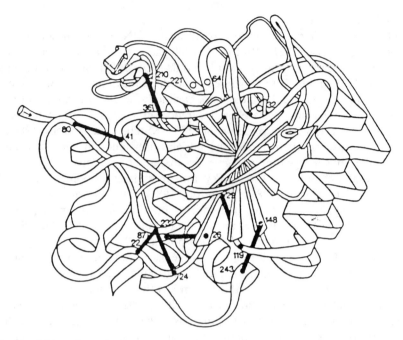

Fig. 6.8—Ribbon diagram of subtilisin with positions of engineered Cys and active site residues (Ser 221, His 65 and Asp 32) labeled with small circles. Disulfide bonds are represented by dark lines. Predicted and observed geometries of these engineered disulfides are shown in Table 6.4. (Redrawn from Mitchinson and Wells, 1989.)

enzymes exhibited essentially the same stability against autolytic inactivation, but the stability of the cross-linked mutant between positions 22 and 87 was less than that of the above enzyme species. Reduction of the disulfide bond lowered the autolytic stability of both oxidized double-Cys mutants relative to their disulfide forms. This corresponds to low autolytic stability for the single mutant with Cys 22 or Cys 87. These observations suggest the H-bond between Thr 22 and Ser 87 in wild-type protein is disrupted on replacing with Cys residue.

Similarly to the work of Wells and Powers (1986), Pantoliano *et al.* (1987) found the candidate pair of Cys residues at positions 22 and 87 which can form a disulfide bond. Pantoliano *et al.* (1987) analyzed the reversible thermal inactivation of the wild-type enzyme and of the oxidized and reduced (dithiothreitol) forms of the double Cys mutant. It is difficult to determine thermodynamic parameters for reversible unfolding with subtilisin by differential scanning calorimetry (DSC), because the enzyme undergoes partial autolytic degradation (Takahashi and Sturtevant, 1981). However, DSC analysis of the three protein samples shows that the value of T_m for the oxidized mutant proteins is apparently higher than that of the reduced mutant and wild-type enzymes (Table 6.3). It was estimated from the values of T_m that the disulfide bond of the oxidized mutant contributes about 1.3 kcal mol^{-1} to the ΔG of unfolding for this protein relative to the reduced form. This value corresponds to less than half of the estimated upper limit ($-\Delta G = 3.2$ kcal mol^{-1}) expected from theoretical considerations (Poland and Scheraga, 1965b), suggesting some deviation

6.2 Mutagenesis with Knowledge of Protein Structure

Table 6.3—T_m values of subtilisin mutants with engineered disulfide bonds. The amino acid residues to be engineered for producing disulfide bond were selected by the following steps. First, the main-chain atoms of two juxtaposed residues were in the same geometrical relationship as the main atoms in known disulfide bonds. Second, the hypothetical sulfur atoms were inserted into structure via computational modeling, and those candidates with unfavorable steric interactions were discarded. Third, the several amino acid pairs identified by this program were subjected to the final requirement that the individual Cys mutations occur at positions that exhibit some variability in evolutionarily related subtilisin primary sequences (Meloun et al., 1985). The positions at 22 and 87 were chosen (see Fig. 6.8). T_m was measured by DSC experiments (Redrawn from Pantoliano et al., 1987)

Variants	T_m(°C)
Wild-type	58.9 ± 0.2
C22(← T)–C87(← S)	62.0 ± 0.1
C22(← T, C87(← S)	56.2 ± 0.1
C87(← S)	57.2 ± 0.2
C22(← T)	56.4 ± 0.1

from an idealized fit. Another possibility is the individual contributions of single amino acids toward the ΔG of unfolding. For example, the single mutant protein in which Thr 22 is replaced with Cys has a lower T_m than the wild-type enzyme (Table 6.3). This lower stability relative to wild-type protein is not due to the oxidation of free Cys in this single mutant enzyme, because its stability does not change in the presence of dithiothreitol. X-ray crystal structures showed that the disulfide bond geometries of the mutant enzymes were different from naturally occurring disulfides (Katz and Kossiakoff, 1986). The thermal inactivation of subtilisin in the presence of urea was also enhanced by introducing the disulfide bond between positions 22 and 87 in the molecule. These observations with the oxidized double Cys mutant were in contrast to the findings by Wells and Powers (1986) with the similar disulfide bond introduced mutant, as described above.

Mitchinson and Wells (1989) then examined the stabilizing effects of engineering disulfide bonds on subtilisin introduced at various sites in the subtilinsin molecule.

They performed model building using coordinates from a 1.8 Å resolution highly refined structure of subtilisin (Bott et al., 1988). Eight candidates selected for engineered disulfides in subtilisin are between positions 22 (Thr) and 87 (Ser) and positions 24 (Ser) and 87 (Wells and Powers, 1986; Pantoliano et al., 1987), and , in addition, between positions 26 (Val) and 232(Ala), positions 29 (Ala) and 119 (Met),

Table 6.4 — Observed and predicted geometries of engineered disulfide bonds in subtilisin[a]

Disulfide	χ_1	χ_2	χ_3	$\chi_{2'}$	$\chi_{1'}$	C_α–C_α distance (Å)	Bond dihedral[b] energy (kcal mol⁻¹)	Structural context
Natural LH[c]		121(112)	−85±9			5.88±0.49	1.7±1.5	
Natural RH[c]		−50(−35)	99±11			5.07±0.73	3.2±1.4	
C22–C87	53(64)	−126(114)	−98(−85)	143(130)	−49(−57)	5.37(5.37)	4.8(4.3)	Loop to loop
C24–C87	−65(−70)	62(61)	96(92)	−171(−152)	−157(−174)	4.59(4.76)	2.5(2.7)	Loop to loop
C26–C232	177(79)	(−151)(134)	−71(115)	−47(86)	−90(−117)	5.57(5.91)	5.0(10.1)	β-sheet to α-helix; buried
C29–C119[d]	−86(−65)	(178)(169)	83(113)	160(144)	−99(−136)	5.88(6.27)	5.4(6.9)	β-sheet to turn; buried
C36–C210	(−81)(−89)		(−90)(92)	(96)(−87)	(−160)(−73)	(5.54)(5.54)	(5.0)(5.6)	Loop to turn
C41–C80[e]	(−175)(−45)	(−10)	(−98)(77)	(126)(111)	(−149)(−76)	(6.30)(6.30)	(5.1)(3.9)	Loop to loop; high-affinity Ca⁺ binding site
C148–C243	(−40)		(−87)	(−63)	(−73)	(5.12)	(3.9)	β-sheet to α-helix

[a] Observed values are from X-ray crystallography (Katz and Kossiakoff, 1986; Katz and Kossiakoff, unpublished results), and predicted values from molecular modeling shown in parentheses. For modeled structures, the C_α–C_α separation is assumed to be fixed from the C_α atoms in the wild-type structure. The dihedral angles (χ_1, χ_2, χ_3, $\chi_{2'}$ and $\chi_{1'}$) are defined as shown in Fig. 2.35.
[b] Calculated from a subroutine of the program AMBER (Weiner et al., 1984).
[c] LH and RH are left-handed and right-handed disulfides, respectively, and these values are average for at least ten natural disulfides (Katz and Kossiakoff, 1986).
[d] C29–C119 modeled by Katz and Mitchinson.
[e] The probability of flexibility in this region of subtilisin in the absence of the structural Ca makes it particularly difficult to judge how close to the actual structure the modeled structure will be. C29–C119 is chosen on the basis of structural homology to the disulfide C34–C123 of proteinase K (Paehler et al., 1984). C148–C243 is chosen because it is similar to the C178–C249 disulfide in proteinase K. The C36–C210 disulfide is introduced to replace the H-bond between the carboxylate of Asp 36 and the main-chain nitrogen of Gly 21. Since Gly 80 is in the Ca binding loop and the carboxylate of Asp 41 is one of the Ca binding ligands (Drenth et al., 1972; Voordouw et al., 1976), C41–C80 was designed to replace the stabilizing effect of the structural Ca. However, the model of the C41–C80 disulfide showed structural adjustment which would be necessary to relieve potential steric hindrance with the side chain of Tyr 214 and main-chain atoms of residues 79 to 81. These designed disulfide bonds connect loops, turns, helices or regions of β-sheets in buried or exposed positions (Fig. 6.8) (Redrawn from Mitchinson and Wells, 1989)

6.2 Mutagenesis with Knowledge of Protein Structure

positions 36 (Asp) and 210 (Pro), positions 41 (Asp) and 80 (Gly) and positions 148 (Val) and 243 (Asn) (Fig. 6.8). The predicted cross-links are at least 10 Å apart from the active center, and exhibited the minimal non-bonded contacts and energetically favorable dihedral angles. Table 6.4 shows the observed and predicted conformational parameters of the mutant proteins.

The mutated subtilisin gene was expressed in *B. subtilis*, and the mutant enzymes were secreted. In almost all double Cys mutant enzymes disulfide bonds were quantitatively formed *in vivo*. All of the single- and double-mutant enzymes had virtually the same specific activity as wild-type enzyme, except for the variant with disulfide bond between positions 148 and 243.

Mitchinson and Wells (1989) measured the half-time for irreversible thermal inactivation at elevated temperatures of the wild-type and mutant subtilisins. As clearly shown in Table 6.5, none of the disulfide mutants are significantly more stable than wild-type enzyme. However, in most cases, formation of disulfide bonds makes a positive contribution to stability compared to their reduced counterparts. The introduction of a single Cys residue either had no effect or was destabilizing, except for the mutant protein in which Met 119 was replaced. They measured the strength of the disulfide bond by the equilibrium constant for reduction of the bond by dithiothreitol. The strength of disulfide varied by over 500-fold in the wild-type and

Table 6.5—Thermal inactivation of subtilisin mutants with engineered disulfide bond at the indicated temperatures with or without dithiothreitol (DTT)

Mutant	half-time (min) for thermal inactivation[a] at					
	55.6 ± 0.2°C		60.6 ± 0.2°C		60.8 ± 0.2°C	
	−DTT	+DTT	−DTT	+DTT	−DTT	+DTT
Wild-type	1500[b]		117	117(85)	83	87(62)
C26(←V)–C232(←A)			121	(93)		
C26(←A)–C119(←M)			75	(51)		
C29(←A)					77	(52)
C119(←M)					149	(126)
C36(←D)–C210(←P)			120	4		
C36(←D)					42	52
C210(←P)					98	93
C148(←V)–C243(←N)	9	102		16		
C148(←V)	191	135		26		
C243(←N)	191	209		34		
C41(←D)–C80(←G)			4[b]	2[b]		
C41(←D)			1[b]			
C80(←G)			4[b]			

[a]The values were obtained from samples inactivated in parallel and always together with a wild-type standard.
[b]This value is approximate because reactions were too slow or too fast to measure half-times accurately
(Redrawn from Mitchinson and Wells, 1989)

mutant enzymes. It was found that no correlation was observed between the strength of the disulfide and its contribution to the stability of subtilisin. Some of these results can be rationalized by destabilizing effects of the introduced Cys residues which disrupt interaction(s) present in the native enzyme structure. It is also possible that the rate of irreversible inactivation depends on the kinetics (the conversion of unfolded state to irreversibly changeed state) and not on the thermodynamic unfolding, so that the entropically stabilizing effect expected from a disulfide bond may not apply.

Dihydrofolate reductase

Villafranca *et al.* (1987) replaced Pro 39 with Cys in dihydrofolate reductase, which possesses a natural Cys 85. Dithionitrobenzoate oxidation of the above mutant enzyme (reduced) yielded a disulfide bonded enzyme between Cys 39 and Cys 85 (oxidized). The disulfide bond was shown to have a geometry close to that of the commonly observed left-handed spiral. Moreover, X-ray crystallographic analysis showed that the conformation of wild-type enzyme was not appreciably affected by the disulfide bond, but the details of the molecule's thermal motion were altered.

Thermodynamic analysis for the reversible unfolding process induced by guanidine hydrochloride showed that the cross-linked enzyme is 1.8 kcal mol^{-1} more stable than either the wild-type or the reduced mutant enzyme. The disulfide bonded enzyme, however, was not more resistant to thermal denaturation. The maximum extent of the unfolding of the cross-linked enzyme by heating was found to be significantly less than that for guanidine unfolding. The thermally unfolded form of the cross-linked enzyme appears to retain some structural integrity in the region near the disulfide bond. Thus the cross-link would not contribute to entropic destabilization of the thermally unfolded form. In addition, the X-ray crystal structure analyses showed

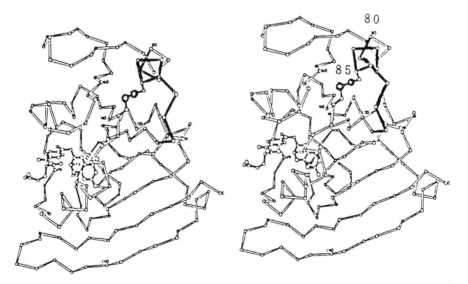

Fig. 6.9—Stereo view of α-carbon backbone of dihydrofolate reductase with a disulfide bond between positions 39 and 85. The chain, residues 80–90, consists of a short α-helix (residues 80–85) and a six-residue loop (residues 85–90). (Redrawn from Villafranca *et al.*, 1987.)

that the flexibility of the region containing residues 80–90 increases on cross-link formation between Cys 39 and Cys 85 (Fig. 6.9). The disulfide bond appears to be too flexible to restrict the flexible motion of the region. This may explain why the disulfide-bonded protein is not more stable against thermal denaturation, relative to the wild-type and reduced mutant enzymes.

6.2.1.2 Other amino acid substitutions

Glycine has no β-carbon and allows more conformational freedom in a polypeptide chain than Ala, which has one methyl group as side chain. In other words the backbone of a Gly residue in solution has greater configurational entropy than Ala. Thus the judicious replacement of Gly to Ala should increase protein stability by reducing the entropy loss on unfolding. Residues such as Thr, Val and Ile with branched β-carbon will also restrict the backbone conformation more than non-branched residues. Similarly, the pyrolidine ring of Pro residue restricts this residue to fewer conformations than are available to the other amino acid residues. Then, many possible amino acid replacements that alter the backbone configurational entropy of unfolding of a protein may be used to increase protein stability.

T4 lysozyme

T4 phage lysozyme has been stabilized in this way by site-directed mutagenesis (Matthews *et al.*, 1987).

X-ray crystallographic analysis of T4 lysozyme showed the two Gly residues at positions 77 and 113 for relevant amino acid substitution, because these residues have conformational angles (ϕ, ψ) that are within the allowed range for amino acids with a β-carbon and could accommodate a β-carbon without interfering with neighboring atoms. Since Ala is conservative and avoids possible secondary effects that might occur with a larger side chain, Gly 77 was replaced with Ala (Fig. 6.10a and b).

A proline residue in a polypeptide chain restricts the (ϕ, ψ) values at the proline itself and limits the (ϕ, ψ) values of the preceding residue. Substitutions of amino acids for Pro must be compatible with these constraints. It was then required that the residue preceding a Pro site has (ϕ, ψ) values within the allowed region, and the values of ϕ and ψ at the substitution site itself were requested to be within the regions $\phi = -50°$ to $-80°$, $\psi = 120°$ to $180°$ or $\phi = -50°$ to $-70°$, $\psi = -10°$ to $-50°$. In addition, the sites where a Pro side chain would sterically interfere with neighboring atoms and where the side chains of a residue appeared to participate in intramolecular interactions within the native structure were removed from consideration. Of the three preferred candidates for Pro substitution thus screened (Lys 60, Ala 82 and Ala 93), the mutant in which Ala 82 is replaced with Pro was constructed (Fig. 6.10c).

The two substitutions (Gly 77 to Ala and Ala 82 to Pro) did not affect enzyme activity. Both substitutions stabilized the enzyme toward reversible and irreversible thermal denaturation at physiological pH. The reversible denaturation of the Ala 77 mutant enzyme was additionally stabilized by $0.4 \text{ kcal mol}^{-1}$. The position 77 is within an α-helical region, so the enhanced stability might be attributed to better helix formation (Menendez-Arias and Argos, 1989). However, entropic stabilization could also be responsible for the increased stability. Thermodynamic parameters for

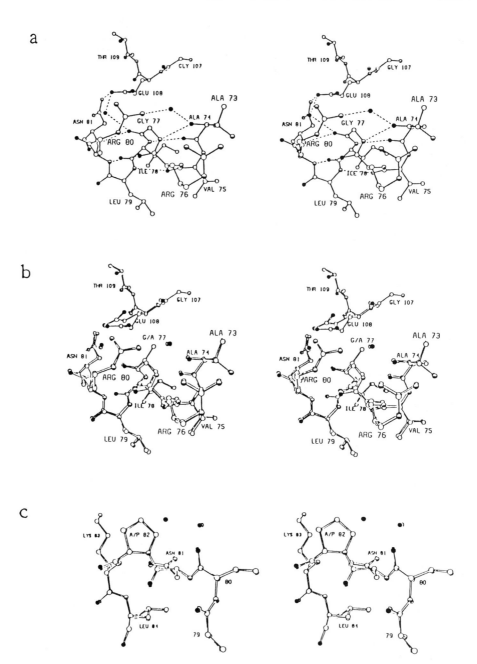

Fig. 6.10—(a) Stereo drawing of the structure of wild-type of T4 lysozyme in the vicinity of Gly 77. Oxygen atoms are solid circles, nitrogen atoms are half-solid circles, and carbon atoms are open circles. H-bonds are shown as broken lines. (b) Superposition of the structures of mutant enzyme with an engineered Ala 77 (←Gly) (open bonds) and wild-type enzyme (solid bonds). (c) Superposition of the structures of mutant enzyme with an engineered Pro 82 (←Ala) (open bonds) and wild-type enzyme (solid bonds). (Redrawn from Matthews *et al.*, 1987.)

reversible thermal denaturation of the wild-type and mutant proteins showed that the increased stability was due to the entropic effect, not to the enthalpy change. High-resolution crystallographic analysis showed that the whole structure of the enzyme was not affected by the substitution, but localized conformational adjustments occurred in the vicinity of the substituted amino acid. The observed increase in thermodynamic stabilization (0.4 kcal mol^{-1}) is about half the value expected from theory (Nemethy *et al.*, 1966). The localized structural change near the substitution site may affect the entropic contribution to the free energy change for unfolding. Appropriately chosen substitution from Gly to Ala will increase stability whether or not the Gly is located within an α-helix.

The structural analysis of the Ala 82 to Pro mutant enzyme showed that its three-dimensional structure is essentially identical with the wild-type enzyme. The

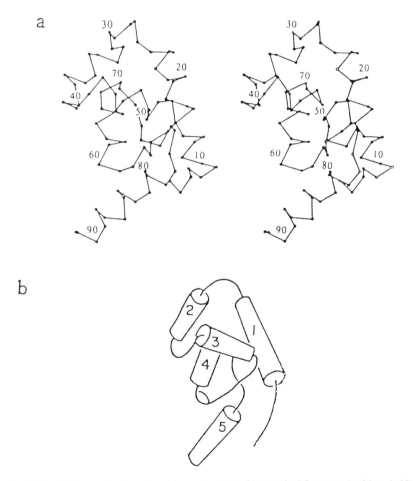

Fig. 6.11—(a) Stereo drawing of backbone structure of N-terminal fragment (residues 4–92) in λ repressor. The fragment contains an extended arm (residues 4–8) and five α-helices (residues 9–23, 33–39, 44–52, 61–69 and 79–92). (b) The helices (1–5) represented by cylinders are arranged from the N-terminus. The N-terminal arm is foreshortened in this view. (Redrawn from Pabo and Lewis, 1982.)

mutant enzyme was stabilized by 0.8 kcal mol^{-1} of free energy, relative to the wild-type enzyme, toward reversible thermal denaturation at physiological pH. This stabilization was also considered to be due to entropy decrease of the unfolded form.

A single amino acid substitution provided a small entropic stabilization, as in this case ($\Delta G = 0.4$–0.8 kcal mol^{-1}). The combination of a number of such amino acid replacements may provide a general strategy for substantial improvement in protein stability (Menendez-Arias and Argos, 1989).

λ repressor

The third α-helix of the N-terminal domain of λ repressor contains two glycines at positions 46 and 48, which are both accessible to solvent (Fig. 6.11) (Pabo and Lewis, 1982). The revertant repressors in which Gly 48 was replaced by Asn or Ser were more stable than the wild-type protein (Hecht *et al.*, 1985). Hecht *et al.* (1986) examined the relationship between α-helical Gly and stability by replacing Gly 46 with Ala and Gly 48 with Ala and by double substituting Gly 46 and 48 by Ala. Ala is one of the best α-helical formers (Chou and Fasman, 1978) and is the most structurally conservative replacement for Gly (Schulz and Schirmer, 1979).

The thermostabilities of the wild-type and mutant proteins were measured by the DSC method. The values of ΔH for thermal denaturation estimated by the present calorimetric method were inaccurate, and the denaturation is not a simple reversible transition between monomeric folded and unfolded states (Hecht *et al.*, 1984). However, it is apparent that both substitutions increase the thermal stability of the N-terminal domain of λ repressor (Table 6.6).

The degrees of contribution to additional thermostability were different between the two single mutations. This may be due to the different effects on protein tertiary structure or sequence-specific effects on helix nucleation or propagation (Hecht *et al.*, 1984). Gly is one of the worst, Ser and Asn are moderate, and Ala is one of the best

Table 6.6—Thermodynamic properties of wild-type and mutant N-terminal domains of λ repressor. ΔH, the enthalpy of denaturation, proportional to the area under the denaturation curve in DSC tracings; $\Delta\Delta G$, free energy change for denaturation of mutant proteins relative to the wild-type protein, which is interpreted as free energies of stabilization relative to the wild-type protein. ΔG for mutants were corrected to T_m for the wild-type protein using the Gibbs–Helmholz equation:

$$\Delta G_{T2}/T2 - \Delta G_{T1}/T1 = \Delta H(1/T2 - 1/T1)$$

assuming the denaturation enthalpy to be independent of temperature (Redrawn from Hecht *et al.*, 1986)

Mutant	T_m(°C)	ΔH (kcal mol^{-1})	$\Delta\Delta G$ (kcal mol^{-1})
G46, G48 (wild-type)	53.4 ± 0.1	65.5 ± 9.8	0
A46(←G), G48	56.5 ± 0.1	70.2 ± 10.5	0.66 ± 0.12
G46, A48(←G)	58.1 ± 0.1	61.6 ± 9.2	0.87 ± 0.13
A46(←G), A48(←G)	59.6 ± 0.1	58.9 ± 8.8	1.10 ± 0.16

6.2 Mutagenesis with Knowledge of Protein Structure

helix-forming residues (Chou and Fasman, 1978). The thermostability of the mutant proteins in which Gly at position 48 was replaced with Ser (or Asn) and Ala increased as the helix-forming ability of the resides increased. This may support the suggestion of the correlation between stability and nucleation or propagation ability of a single α-helix.

6.2.2 Altering stabilizing factors

Another fundamental strategy for stabilizing proteins by site-directed mutagensis is to increase stabilizing factors (ΔH) in the folded protein. As in the case of protein stabilization by reducing destabilizing factors (ΔS), the strategy for enhancing stabilizing factors is based on information on the X-ray crystal structure of the protein to be engineered.

6.2.2.1 *Internal hydrophobicity*

Protein folding is driven by the hydrophobic properties of non-polar amino acid residues (Schulz and Schirmer, 1979). The intramolecular hydrophobicity will then contribute to the stability of proteins (Rose *et al.*, 1985; Eisenberg and McLachland, 1986; Baldwin, 1986). Strengthening the internal hydrophobicity by amino acid substitution may increase protein stability.

T4 Lysozyme

The side chain of Ile 3 in T4 lysozyme contributes to the major hydrophobic core of C-terminal lobe and also helps to link the N- and C-terminal domains (Remington *et al.*, 1978; see also Fig. 6.2). The side chain is about 80% inaccessible to solvent and contacts the side chains of Met 6, Leu 7 and Ile 100 as well as the main chain of Cys 97 (Fig. 6.12). These four residues are buried within the protein interior. It is suggested that the hydrophobicity of the residue at position 3 contributes to the stability of the molecule. The effect of replacement of Ile 3 with various amino acids on the stability of the protein was then examined by Matsumura *et al.* (1988a).

The stability of the wild-type and mutant proteins was measured with thermal reversible denaturation at pH 2.0. Fig. 6.13(a) shows the correlation between residue hydrophobicity (ΔG_{tr}) and increased stability ($\Delta \Delta G$) of various enzyme species. There is a clear correlation between ΔG_{tr} and $\Delta \Delta G$, except for mutant proteins with Phe, Tyr, Trp and Cys (disulfide). The data on other proteins display an essentially linear relationship, and the slope of the straight line is about 0.8. This indicates that 1 kcal mol^{-1} of transfer free energy from water to ethanol contributes, on average, about 0.8 kcal mol^{-1} to the overall stability of the protein.

One of the characteristics of the hydrophobic interaction is that it is independent of pH. The plots representing the correlation between the values of $\Delta \Delta G$ at pH 2.0 and 6.5 for most mutant proteins were linear (Fig. 6.13b). This shows that the change in protein stability on amino acid substitution at position 3, as measured by the change in ΔG, is independent of pH. Hydrophobic interaction may be the major factor governing protein stabilization at position 3.

Two of the greatest departures from linearity occur for the mutant proteins substituted with Asp and Glu. The loss of protein stability at pH 6.5 relative to pH 2.0

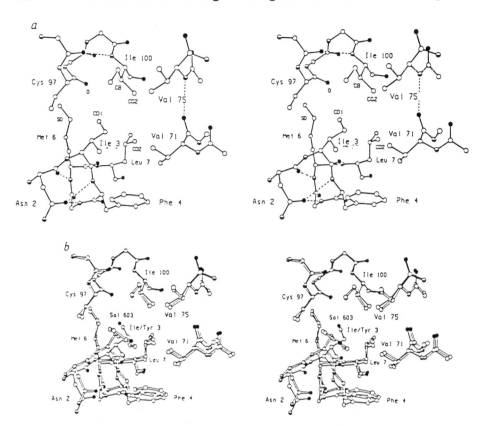

Fig. 6.12—(a) Stereo drawing of T4 lysozyme in the vicinity of Ile 3. Residues 3–7 lie in one α-helix, residues 97–100 are within another helix, and residues 71–75 form part of a third helix. Those atoms that are within 4 Å of the side chain of Ile 3 are labeled individually. The side chain if Ile 3 is about 80% inaccessible to solvent and contacts the side chains of Met 6, Leu 7 and Ile 100 as well as the main chain of Cys 97. These four residues are buried within the protein interior. Carbon atoms are shown as open circles, nitrogen atoms are half-filled and oxygen atoms are drawn solid. (b) Superposition of the structures of the mutant enzyme with Tyr 3 (←Ile) (open bonds) and wild-type enzyme (solid bonds). The side chain of Tyr is not accommodated within the interior of the protein but rotates about 120° so that it is largely exposed to the solvent, creating a cavity which is partly occupied by a water molecule (Sol 603 in the figure). The 0.7–1.7 Å movement of α-, β- and γ-carbon atoms of Tyr 3 into the hydrophobic cavity and the associated shifts (0.6 to 1.1 Å) of the N-terminal chain (residues 1–9) are observed.
(Redrawn from Matsumura *et al.*, 1988a.)

for the two proteins are consistent with the decreased hydrophobicities of these residues. The residues of Asp and Glu are ionized at pH 6.5, but not at pH 2.0, and have decreased hydrophobicities at pH 6.5. The values for ΔG_{tr} of the ionized Asp and Glu are negative, while those for almost all other amino acid residues are positive.

As can be seen in Fig. 6.13(b), the Tyr mutant, and also Phe and Trp, diverge from linearity. The reason is apparent from the structure of the mutant enzyme, as compared to the wild-type protein. The conformation around position 3 of the Tyr mutant is shown in Fig. 6.12. The side chain of the Tyr protrudes into the solvent and remains partially solvated in the folded form, and does not exhibit its full

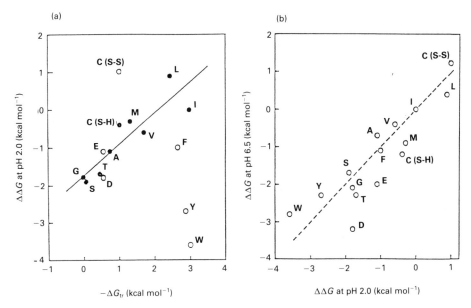

Fig. 6.13—(a) Free energy of stabilization ($\Delta\Delta G$) of mutant lysozymes at pH 2.0, plotted against the free energy of transfer ($-\Delta G_{tr}$) of individual amino acids from water to ethanol (Tanford, 1962). Protein free energies are plotted relative to wild-type with Ile and transfer free energies are relative to Gly. (b) Free energy of stabilization of mutant enzymes at pH 6.5 plotted against stabilization at pH 2.0. Cys (S—S) means the disulfide formation between natural Cys 97 and mutant Cys 3. (Redrawn from Matsumura et al., 1988a.)

hydrophobic potential. The oxidized form of the Cys mutant exhibits more enhanced stability, as compared to the reduced form and wild-type protein, and is presumably due to entropic effects (see section 6.2.1.1).

The accessible surface area (Lee and Richards, 1971a; see also Chapter 2) was correlated with the energy of transfer from water to organic solvents (Chothia, 1974). The changes in the surface area that occur during protein folding were then used to calculate the contribution of hydrophobic interactions to the free energy of folding.

The solvent-accessible surface area of natural Ile and replaced Val, Ala and Gly at position 3 were calculated according to Lee and Richards (1971), and the free energy changes of the Ile, Val and Ala relative to Gly on protein folding were calculated according to Chothia (1974) and are shown in Table 6.7 (Matsumura et al., 1988a). The differences in calculated free energy change with the wild-type (Ile) and mutant side chains (Val and Ala), relative to Gly on transfer from a fully solvated form to a completely buried form, are in good agreement with the differences in free energy change for unfolding of the wild-type (Ile) and Val and Ala mutant proteins, relative to the Gly-replaced mutant protein. Fig. 6.13 and Table 6.7 show that the local hydrophobic effect at a specific site contributes to the overall stability of the protein.

Ribonuclease

Ribonuclease from *Bacillus amyloliquefaciens* (barnase) is a monomer of 110 amino acid residues of molecular weight 12 382, which is much smaller than T4 lysozyme

Table 6.7 — Contribution of hydrophobic free energy to stabilization of T4 lysozyme on transferring amino acid residue at position 3 from a fully solvated form to the interior of the unfolded form. Solvent-accessible surface area was calculated by the method of Lee and Richards (1971). A_0 is the solvent-accessible surface area of the residue in an extended ($\varphi = -139°$, $\psi = 135°$) tripeptide Asn–X–Phe. A is the area of residue 3 in the folded protein. The differences in solvent exposure ($A_0 - A$) is assumed to correspond to the area that is buried when protein folds; e.g. ($A_0 - A$) for Gly is the value of ($A_0 - A$) after subtraction on the accessible surface area of Gly (39.7 Å^2). Coordinates for calculation of A were from crystal structures in the case of Ile and Val and were inferred from model building in other cases. $\Delta\Delta G$, the calculated free energy of folding relative to Gly, is obtained by multiplying the previous column by 24 cal Å^2 mol^{-1}/ (Chothia, 1974). $\Delta\Delta G$ is the difference between the free energy of unfolding of the mutant proteins and that of wild type at T_m of wild-type protein (Redrawn from Matsumura et al., 1988a).

Amino acid residue at position 3	A_0 (Å^2)	A (Å^2)	$A_0 - A$ (Å^2)	$A_0 - A$ (Gly) (Å^2)	$\Delta\Delta G_{calc}$ (kcal mol^{-1})	$\Delta\Delta G_{obs}$ (kcal mol^{-1})
Ile	151.6	20.5	131.1	91.4	2.2	1.8–2.1
Val	128.7	17.6	111.1	71.4	1.7	1.2–1.7
Ala	87.1	14.8	72.3	32.6	0.8	0.7–1.4
Gly	56.3	16.6	39.7	0.0	0.0	0.0

(mol. wt = 18 720) and subtilisin BPN' (mol. wt = 27 537). In barnase the α-helix formed from residues 6–18 packs against the antiparallel β-sheet formed from residues 50–55, 70–75, 85–91, 94–101 and 106–108 (Mauguen et al., 1982). Barnase is composed of one domain and undergoes a reversible thermal or urea-induced denaturation. This enables thermodynamic measurements to be made on the factors that influence protein unfolding and stability.

The mutant enzymes in which Ile 96 is substituted with Val, Ile 96 with Ala and Phe 7 with Leu were constructed (Kellis et al., 1988). These mutations remove hydrophobic interactions where the helix docks on the β-sheet, effectively introducing cavities into the enzyme. In order to measure the change in free energy of the reversible denaturation of the enzyme, urea-induced denaturation was chosen instead of thermal denaturation. The reason was first that the thermal denaturation is not adequate for quantitative comparison of different proteins with different values of T_m (Creighton, 1983). The enthalpy of unfolding of proteins varies with temperature and the van't Hoff plots are not linear. Thermodynamic quantities are consequently not easily extrapolated to different temperatures. Urea denaturation, on the other hand, may be performed at a single temperature for a whole series of mutants. Second, urea denaturation gives a more completely unfolded protein than does thermal denaturation (Creighton, 1983; Pace, 1986).

The free energy of unfolding in the presence of urea (ΔG) was conventionally estimated by extrapolating the free energy of unfolding at each individual concentration of urea to zero concentration, assuming that they are linearly related (Pace, 1986). The values of ΔG for the wild-type and Val, Ala and Leu mutant enzymes were 10.5 ± 0.2, 9.3 ± 0.2, 6.5 ± 0.1 and 6.44 ± 0.04 kcal mol^{-1}, respectively. The results indicate that these amino acid substitutions destabilize barnase. The values of $\Delta\Delta G$ ($\Delta G_{wild} - \Delta G_{mutant}$) for the mutations from Ile to Val and from Ile to Ala are 1.1 and 4.0 kcal mol^{-1}, respectively. Creation of a cavity the size of a —CH$_2$— group destabilizes the enzyme by 1.1 kcal mol^{-1}. Such experiments will give direct measurements of the destabilizing effects of cavities in proteins, which will demonstrate the way to stabilize protein structures: make substitutions to fill in holes that occur in the native structure.

6.2.2.2 Electrostatic interactions

Addition or strengthening of non-covalent interactions other than hydrophobic interactions has been considered as a way of stabilizing proteins (Nosoh and Sekiguchi, 1988). As described in Chapter 4 additional electrostatic interactions have been suggested to contribute to increased thermostability of thermophilic proteins as compared to their mesophilic counterparts (Colman *et al.*, 1972; Perutz and Raidt, 1975; Perutz, 1978; Argos *et al.*, 1979; Walker *et al.*, 1980; Ruegg *et al.*, 1982).

T4 lysozyme

Negatively charged amino acid residues are very frequently observed near the N-terminals of helices in protein structures (Richardson and Richardson, 1988). The thermostability of semi-synthetic ribonuclease was affected by changes in charge at the ends of the helix (Mitchinson and Baldwin, 1986). These indicate a possible introduction of charged residues at sites designed to interact with the positive charge attributed to the α-helix dipoles (Hol *et al.*, 1978). Nicholson *et al.* (1988) searched for the candidate pairs of α-helix and amino acid substitution site in the structure of T4 lysozyme. Ser 38 and the helix, residues 39–50, and Asn 144 and the helix, residues 143–155, were selected, because the helices have no negatively charged groups near their N-terminals, and they are not H-bonded with other helices. The introduction of a charged residue would be counterproductive. The locations of the helix and substitution site are shown in Fig. 6.14. Ser 38 and Asn 144 were both replaced with Asp.

The activities of the single mutants in which Ser 38 and Asn 144 are respectively replaced with Asp, and the double mutant in which both Ser 38 and Asp 144 are replaced with Asp, were higher than that of the wild-type protein: 320%, 160% and 430% that of the native enzyme, respectively. The values of T_m as monitored by the changes in circular dichroism (CD) at 223 nm of the single mutants were about 2°C higher than the wild-type at pH 3.5. The double mutant showed an additive effect with about 4°C increase over the wild-type's melting temperature, which corresponds to an increase of 1.6 kcal mol^{-1} in the free energy of stabilization of the double mutant relative to the wild-type enzyme. Such additivity in stabilization resulting from multiple mutations has been reported by others (Hecht *et al.*, 1986; Matsumura *et al.*, 1986b). Here again, the combination of several mutations, each of which may cause only a small increase in stability, can provide substantial improvement in protein stability (Matthews *et al.*, 1987; see also section 6.2.1.2).

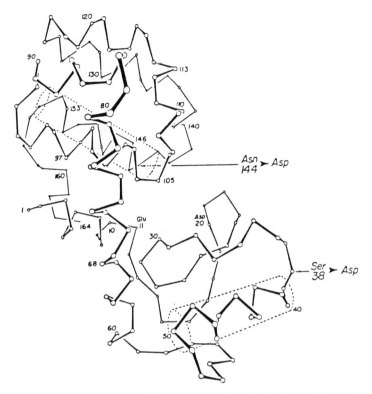

Fig. 6.14—Backbone structure of T4 lysozyme showing the locations of the amino acid substitution. Of 11 α-helices in T4 lysozyme, the helix B (residues 39–50) and J (residues 143–155) were chosen. The helices are outlined with a broken line. (Redrawn from Nicholson et al., 1988.)

From the comparison of X-ray crystal structures of the single mutants and wild-type enzymes, it was confirmed that mutant structures are stabilized by an electrostatic interaction between the negatively charged side chain and the positive charge attributed to the α-helix dipole. This may be strongly supported by the finding that it was only at pH values where the aspartates are negatively charged in the folded state that the mutant structures are more stable than the wild-type.

Subtilisin BPN′

Computer-modeling of subtilisin BPN′ revealed ten candidates for amino acid substitutions which lead to potential formation of electrostatic interactions, generally on the surface of the protein molecule and not close to the active site (Erwin et al., 1990). Substitutions to replace the candidate amino acid residues with Glu or Arg by site-directed mutagenesis were attempted, because Glu and Arg replacements occur frequently from mesophiles to thermophiles (Menendez-Arias and Argos, 1989). Nine single and one double mutated enzymes thus constructed exhibited only slightly modified activities expressed by K_m and k_{cat}. The variations in pI of the mutated enzymes are as predicted from the amino acid replacements.

Irreversible thermal inactivation experiments which reflect autolytic stability

(Wells and Powers, 1986; Pantoliano et al., 1987) showed that the double mutant enzyme, Gln 19 to Glu and Gln 271 to Glu substitutions, was most dramatically increased where a twofold difference over wild-type was seen. The mutant in which Asn 77 is replaced with Lys showed a 30% increase in stability. Two single variants, Val 51 to Lys and The 164 to Arg, were considerably less than the wild-type. The stability of other mutants was almost the same as that of wild-type.

Relative thermodynamic stabilities of the wild-type and mutant proteins compared the values of T_m determined by DSC at pH 8 or at pH 10. The single mutant with Gln 19 to Glu substitution had a T_m of 1.6°C higher than the wild-type, but the single mutant with Gln 271 to Glu substitution and double mutant with Gln 19 and 271 to Glu substitutions had values 1.5 and 1.2°C below the wild-type, respectively. X-ray crystallographic analysis of the double mutant revealed that Glu 19 forms a salt bridge with Lys 237, whereas Glu 271 does not form one with its putative partner, Lys 12. At pH 10, the mutant with Glu 19 had a T_m of 71.5°C and that of the wild-type was 71.1°C. This suggests that the salt bridge formed in the single variant is ruptured due to the deprotonation of Lys 237. In the case of the double mutant, the value of T_m may reflect a balance between the stabilizing effect of the Glu 19 variant and the destabilizing effect of the Glu 271 variant.

6.2.2.3 *Hydrogen bonds*

A great number of studies on structural comparison of thermophilic and mesophilic proteins have so far been made (Chapter 4). However, none has been reported on a positive contribution of H-bond to enhanced stability of thermophilic proteins as compared to their mesophilic counterparts.

Thr at position 157 in T4 lysozyme is known to be located in an irregular loop on the surface of the protein (Weaver and Matthews, 1987). The natural mutant in which Thr 157 is replaced with Ile is temperature sensitive (Grutter et al., 1987). X-ray crystallographic analysis showed that the hydroxyl group of Thr 157 is H-bonded to the main chain amide of Asp 159 in the wild-type enzyme, but the H-bond is ruptured in the mutant protein (Fig. 6.15). The replacement of Thr 157 with 12 different amino acids by site-directed mutagenesis was then attempted, to reveal the relation of the local structure around the position 157 to stability of T4 lysozyme (Alber et al., 1987).

The thermodynamic stability of the wild-type and mutant enzymes was assessed by measuring the molar ellipicity at 223 nm in the CD spectrum of a solution of the purified protein at different temperatures. From the measurements the values of T_m and ΔG were estimated. The change in T_m caused by amino acid substitution at position 157 and the corresponding change in the free energy of stabilization is shown in Table 6.8. None of the 13 mutant proteins were more stable than the wild-type.

Detailed X-ray crystallographic analysis revealed the relation of the structure, especially around position 157, to stability of the wild-type and mutant proteins. The most remarkable pattern is that the buried main chain amide of Asp 159 is H-bonded in the most stable variants with Thr (wild-type), Asn, Ser, Asp and Gly at position 157. The loss of this H-bond in the folded state is the major cause for loss of stability. However, different H-bonds are associated with different net contributions to stability, as shown in Fig. 6.16 and Table 6.9, although it is not clear that their contribution

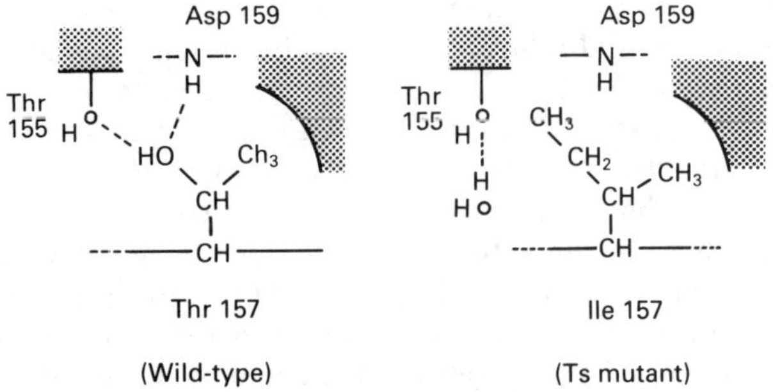

Fig. 6.15—Schematic illustration comparing the local environment of Thr 157 in wild-type lysozyme and Ile 157 in temperature-sensitive mutant. Thr 157 is located in an irregular loop on the surface of the protein (see Fig. 6.14). The replacement of the hydroxyl group of Thr with an ethyl group of Ile leads to a complex, though localized, structural change. H-bonds to the Thr 157 hydroxyl group are lost in the mutant protein, but tighter binding of a water molecule in the vicinity of the Thr 157 hydroxyl group may partially compensate for this change. The β-carbon of residue 157 shifts about 0.7 Å, and a rotation about the C_α—C_β bond causes a change in the contacts of the side chain γ-methyl group. The side chain of Asp 159 moves 1.1 Å to accommodate the bulkier ethyl group at position 157. See also Table 6.9. (Redrawn from Alber et al., 1987a.)

Table 6.8—Thermodynamic stability of T4 lysozyme with different amino acids at position 157. ΔT_m is the change of T_m of mutant proteins relative to wild-type protein (T_m is 42°C at pH 2.0). To ensure reversibility, unfolding and refolding were both monitored by CD (Redrawn from Alber et al., 1987)

Amino acid at position 157	ΔT_m(°C)	$\Delta\Delta G$ (kcal mol^{-1})
Thr (wild-type)	—	—
Asn	−1.7	0.45
Ser	−2.5	0.66
Asp	−4.2	1.1
Gly	−4.2	1.1
Cys	−4.9	1.3
Leu	−5.0	1.3
Arg	−5.1	1.3
Ala	−5.4	1.4
Glu	−5.8	1.5
Val	−6.0	1.6
His	−7.9	2.1
Phe	−9.2	2.4
Ile (ts mutant)	−11.0	2.9

6.2 Mutagenesis with Knowledge of Protein Structure

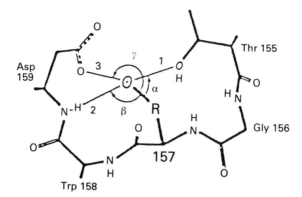

Fig. 6.16—Schematic diagram showing the geometry of the H-bond network formed with different amino acids at position 157 in T4 lysozyme. The three potential H-bond lengths (1, 2, 3) and angles (α, β, γ) are listed in Table 6.9. H-bonds to the amide of Asp 159 are different in detail with Asn, Asp, Ser and Gly at position 157 (see also Table 6.9). These or other differences in the folded and unfolded states of these mutants could account for the observed small differences in stability (Table 6.8). (Redrawn from Alber et al., 1987a.)

Table 6.9—Geometry of the H-bond network with different amino acids at position 157. Interatomic distances 1, 2, 3 are the three potential H-bond lengths, and their angles are given by α, β, γ (see Fig. 6.16). Asp (1) is the conformation of Asp 157 that resembles Asn 157. Also induced is the crystallographic thermal factor, B, of the central H-bonded atom (Redrawn from Alber et al., 1987a)

Amino acid	Atom	Interatomic distance (Å)			Angle (°)			B (Å^2)
		1	2	3	α	β	γ	
Thr	O^γ	2.8	3.2	3.2	119	103	106	20
Ser	O^γ	2.8	3.4	3.2	119	107	114	30
Asn	O^δ	3.1	3.1	3.2	97	130	101	30
Asp(1)	O^δ	2.9	3.1	3.1	100	125	94	—
Gly	H_2O	2.9	3.1	2.8	—	—	—	29

to stability could compensate for the loss of the H-bond between Thr 157 and Asp 159 in the mutant protein.

The stability of wild-type protein could not be sufficiently accounted for by the H-bounds. This is shown by comparison of the proteins with Thr and Ser at position 157. In spite of the similar conformations of the proteins, the stability of wild-type protein is reduced by replacement of Thr with Ser. This may be partly explained by the contribution of the γ-methyl group of Thr. This contribution could be a direct effect, such as the increase of van der Waals contact of burying hydrophobic surface area in the folded state, or an indirect effect, e.g. by decreasing the flexibility of the unfolded state or increasing the rigidity of H-bonds to the γ-hydroxyl group. The difference in stability between the Ile and Val mutant proteins may also be accounted for by the hydrophobic effect of the δ-methyl group of Ile.

6.2.3 Compensating stabilizing–destabilizing factors

To fully understand the mechanism of marginal stability of protein folded state, the effects of amino acid substitution on the structural change on mutation should be examined by X-ray crystal analysis; effects on the alteration of thermodynamics for reversible denaturation should also be examined, as already shown by the above mutation examples.

A large collection of mutants of staphylococcal nuclease have been collected (Shortle and Lin, 1985). The enzyme undergoes two-state reversible denaturation (Shortle, 1986). A study on over 40 different mutant proteins by guanidine hydrochloride denaturation has revealed that amino acid substitutions affected both the free energy of denaturation (ΔG) and the rate of ΔG change as a function of denaturant ($\Delta G/dC$) (Shortle and Meeker, 1986). The altered values of $d(\Delta G/dC)$ were explained as a manifestation of large changes in chain conformation or chain–solvent interactions which occur in the denatured state.

To understand more fully the physical mechanism of reversible denaturation, the mutants were analyzed by thermal denaturation (Shortle *et al.*, 1988). Analysis was performed by monitoring the change of tryptophan fluorescence of protein. It was confirmed that the denaturation processes of wild-type and mutant proteins could be monitered by this analytical procedure. The thermodynamic parameters estimated for reversible thermal denaturation of wild-type and mutant enzymes are listed in Table 6.10. Some mutations stabilized and others destabilized staphylococcal nuclease, as judged from the values of T_m and ΔG. The values of ΔH and ΔS for destabilized mutant proteins were both greater than those for the wild-type and stabilized mutant proteins.

Such $\Delta H/\Delta S$ compensation was reported with a small molecule reaction in solution and with a number of reactions involving the binding of ligands to proteins (Lumry and Rajender, 1970). For several single amino acid substituted mutants of T4 lysozyme (Hawkes *et al.*, 1984), the reduction in ΔH for denaturation at 46.9°C was compensated by the reduction in ΔS, which resulted in only a small change in ΔG. For example, for the mutant enzyme in which Arg 96 is replaced by Thr, the value of ΔH was reduced by 27% and ΔS by 25%.

According to Lumry and Rajender (1970), the protein unfolding

$$N(\text{native}) \rightleftharpoons U(\text{unfolding})$$

is coupled to a reaction between the energetically different state of water

$$n(H_2O)_N \rightleftharpoons n(H_2O)_U$$

Thus, the overall ΔH and ΔS for denaturation are

$$\Delta H = \Delta H_{N \to U} + n\Delta H_{WN \to WU}$$

$$\Delta S = \Delta S_{N \to U} + n\Delta S_{WN \to WU}$$

6.2 Mutagenesis with Knowledge of Protein Structure

Table 6.10—Thermodynamic parameters of the reversible denaturation reactions of staphylococcal nuclease at pH 7.0[a]. The graphs of relative Trp fluorescence versus temperature were depicted, and thermodynamic parameters were then obtained by these denaturation curves. (Redrawn from Shortle et al., 1988)

Nuclease	ΔG(321 K)	ΔH(321 K)[b]	ΔS(321 K)	ΔG(293 K)[c]	ΔG(293 K)[d]
Wild-type	+1.4	73.3	0.224	+6.1	+5.6
L66($\leftarrow$V)	+1.8	58.5	0.177	+5.6	+5.4
V88($\leftarrow$G)	+1.7	53.7	0.162	+5.2	+4.6
L66($\leftarrow$V), V88($\leftarrow$G)	+1.6	41.6	0.125	+3.9	+3.5
L66($\leftarrow$V), S79($\leftarrow$G), V88($\leftarrow$G)	+0.6	37.2	0.114	+3.2	+2.6
T69($\leftarrow$A)	−1.4	74.5	0.236	+3.3	+2.9
M18($\leftarrow$I), S90($\leftarrow$A)	−1.6	80.2	0.256	+2.7	+2.7

[a] Values of ΔG and ΔH are in kcal mol^{-1}, and those of ΔS are in kcal mol^{-1} K^{-1}.
[b] Estimated by fitting the ΔH data showing the graph of ΔH versus T_m to a second-order polynomial equation in T and solving at 47.9°C.
[c] Estimated from ΔH (321 K), ΔS (321 K), and ΔC_p (25–42°C).
[d] Estimated from guanidine hydrochloride denaturation at 293 K and linear extrapolation to zero denaturant.

If the number of water molecules n that take part in this reaction greatly increases or decreases as a result of conformational changes in the structure of the reactant; i.e. the folded and unfolded forms of a protein, the contribution of water molecules to ΔH and ΔS will greatly increase or decrease in parallel, resulting in $\Delta H/\Delta S$ compensation.

For guanidine hydrochloride denaturation of nuclease, the values of ΔC_p (change in heat capacity for protein unfolding) in the temperature range 25–42°C were correlated to the values d(ΔG)/dC at 30°C (C is the denaturation concentration). The value of d(ΔG)/dC is proportional to the increase in solvent-accessible surface area (ΔA) on protein unfolding (Schellman, 1978). Since the ΔC_p comes from the additional water of hydrophobic hydration bound to the unfolded state (Privalov, 1979), the correlation between ΔC_p and d(ΔG)/dC could be a consequence of changes in ΔA caused by the amino acid substitution. One of the probable explanations for the striking $\Delta H/\Delta S$ compensation may be due to major alterations in the hydration of the denatured state resulting from the amino acid substitutions. The large changes in d(ΔG)/dC for gunanidine hydrochloride denaturation of the nuclease mutants have already been attributed to the U state rather than the N state (Shortle and Meeker, 1986).

The results indicate the importance of thermodynamic treatments in estimating the stability change on site-directed mutagenesis of proteins, and the important role of water in the energetics of protein unfolding and stability.

6.2.4 Protection against chemical destibilization

Irreversible inactivation, especially oxidative, of some enzymes is triggerd by chemical damage of amino acid residues, such as deamidation of Asn residues which is followed by hydrolysis of the peptide bond at the Asp residues, destruction of Met residues and cleavage at the β-position of Cys residues (Brot and Weissbach, 1983; Ahern and Klibanov, 1985). Intramolecular thiol/disulfide interchange also causes irreversible inactivation (Wang *et al.*, 1984). Substitution for Asn, Met and Cys, therefore, might protect from such incactivation.

Protection of Met

Met residue 222 in *B. amyloliquefaciens* subtilisin (Fig. 6.17) is considered to be converted to methionine sulfoxide by H_2O_2, leading to irreversible inactivation (Stauffer and Eston, 1969). The residue was replaced with another 19 amino acids by site-directed mutagenesis, and the residual activity of wild-type and mutant enzymes was measured after exposure to 0.1 or 1.0 M H_2O_2 (Estell *et al.*, 1985). All mutant proteins, except that replaced with Cys, exhibited less specific activity than that of wild-type, 0.3–53% relative to the wild-type activity. The specific activity of the Cys mutant was 138% of wild-type. In general, small amino acids were the most active, followed by the amino acids with amides and aliphatic side chains. Bulky aromatic and charged amino acid substitutions were less active. The kinetic constants for some mutants and wild-type enzymes are shown in Table 6.11.

The stability against H_2O_2 (0.1 and 1.0 M) inactivation of wild-type and some mutant proteins is shown in Fig. 6.18. The mutants containing non-oxidizable amino acid residues, i.e. Ser, Ala and Leu, were resistant to inactivation even by 1.0 M H_2O_2, whereas the mutant with oxidizable residues Met and Cys were readily inactivated.

Protection of Asn

In contrast to reversible thermal denaturation, the unfolded state is thermally converted to denaturation products in an irreversible denaturation process of most

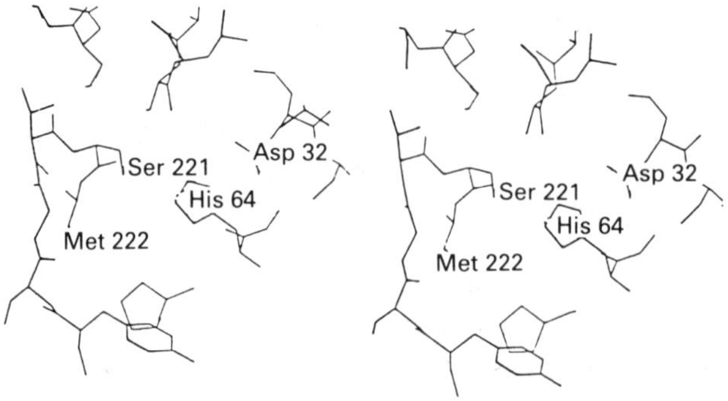

Fig. 6.17—Stereo view of the structure of the active site of subtilisin from *B. amyloliquefaciens* (Wright *et al.*, 1969). Ser 221, His 64 and Asp 32 form the catalytic trial typical of serine protease. Met is the residue identified to be oxidized to the sulfoxide by H_2O_2 accompanied by inactivation (Stauffer and Eston, 1969). (Redrawn from Estell *et al.*, 1985.)

6.2 Mutagenesis with Knowledge of Protein Structure

Table 6.11 — Kinetic constants for subtilisin mutants with different amino acids at position 222. Enzyme was assayed at 25°C and at pH 8.6, and K_m and V_{max} were determined from initial rate measurement. The value of k_{cat} was calculated from the relationship of $k_{cat} = V_{max}/[\text{enzyme}]$ (Redrawn from Estell et al., 1985)

Replacement	k_{cat} (s^{-1})	K_m (M)	k_{cat}/K_m (M^{-1} s^{-1})
Met (wild-type)	50 ($\pm$1)	1.4($\pm$0.05) $\times$ 10^{-4}	36 $\times$ 10^4
Cys	84 ($\pm$2)	4.8($\pm$0.3) $\times$ 10^{-4}	20 $\times$ 10^4
Ser	27 ($\pm$1.8)	6.3($\pm$0.6) $\times$ 10^{-4}	4 $\times$ 10^4
Ala	40 ($\pm$1)	7.3($\pm$0.4) $\times$ 10^{-4}	5 $\times$ 10^4
Leu	5 ($\pm$0.1)	2.6($\pm$0.2) $\times$ 10^{-4}	2 $\times$ 10^4

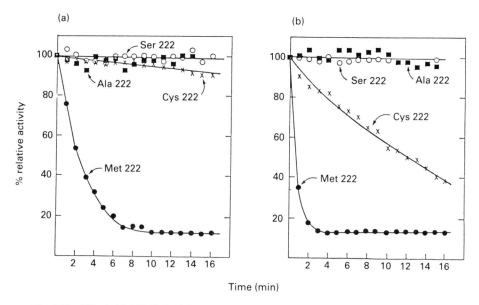

Fig. 6.18 — Effect of 0.1 M H_2O_2 (a) or 1 M H_2O_2 (b) on the activity of wild-type and mutant subtilisin. Met 222 in wild-type enzyme was replaced with Ser, Ala and Cys. (Redrawn from Estell et al., 1985.)

proteins, by aggregation of the unfolded form, or by conformational processes which result in an incorrectly folded form, or by covalent processes which cause cleavage of peptide bonds (Klibanov, 1983).

The processes causing irreversible inactivation of hen egg-white lysozyme at 100°C were found to be deamidation of Asn residues, hydrolysis of peptide bonds at the Asp residues, destruction of disulfide bonds, and formation of incorrect structures from the unfolded state (Ahern and Klibanov, 1985).

On treatment at pH and 100°C, the thermal inactivation of lysozyme was caused by the deamidation of Asn residues followed by the cleavage of peptide bonds at the Asp acid (Table 6.12). The deamidation of Asn residues was the only cause for the thermal inactivation of the enzyme at pH 6. At pH 8, on the other hand, the cause of the irreversible inactivation was a combination of deamidation of Asn, hydrolysis of peptide bonds at Asp, and destruction of disulfide bonds.

Yeast triose-phosphate isomerase consists of two identical subunits; each subunit contains 12 Asn residues, and three of them are at the interfacial van der Waals surface: Asn 14, 65 and 78 (Ahern et al., 1987). Stereo drawings showing the Asn residues in the polypeptide chain and the Asn residues at the subunit interface are represented in Fig. 6.19.

On exposing at 100°C and at pH 6, the enzyme rapidly aggregates and forms a precipitate, which is an inherently ill-defined process that obscures understanding of the mechanism of protein inactivation. When the protein is briefly heated in the presence of guanidine hydrochloride and then cooled and diluted, activity is completely recovered. On longer exposure to 100°C, however, the extent of such activity recovery decreased. Such irreversible change of the enzyme was shown to be due to the deamidation of Asn. Deamidation of the three Asn at positions 14, 65 and 78 located at the interfacial van der Waals surface results in the formation of charged

Table 6.12—Rate constants of irreversible inactivation of hen egg white lysozyme at 100°C. Enzyme sample was incubated at 100°C and at various pHs, and portions were periodically withdrawn and analyzed as follows. *Assayed for the enzyme activity. @Determined by non-equilibrium polyacrylamide gel electrophoresis. ‡Determined by SDS gel electrophoresis of the reduced enzyme; it is assumed that hydrolysis of a single peptide bond results in enzyme inactivation. §Determined by titration of SH groups formed on reduction of cystine residues; it is assumed that destruction of any cystine residue results in enzyme inactivation. ∥Determined as the difference between the time course of irreversible thermoinactivation of lysozyme in the absence and in the presence of 8 M acetamide (Redrawn from Ahern and Klibanov 1985)

Irreversible thermoinactivation	Rate constant (h^{-1})		
	pH 4	pH 6	pH 8
Directly measured overall process*	0.49	4.1	50
Due to individual mechanisms			
Deamidation of Asn residues@	0.45	4.1	18
Hydrolysis of Asp-X peptide bonds‡	0.12	0	0
Destruction of cystine residues§	0	0	6
Formation of incorrect structures∥	0	0	32

Ch. 6]   **6.2 Mutagenesis with Knowledge of Protein Structure**   175

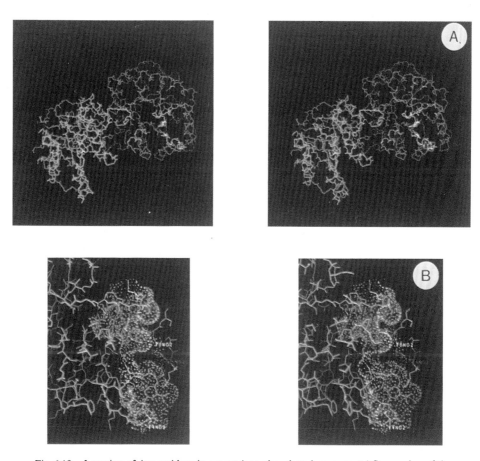

Fig. 6.19—Location of Asn residues in yeast triose-phosphate isomerase. (a) Stereo view of the polypeptide chains (blue and purple) and the Asn residues (green and red) of the two identical subunits. (b) Stereo view showing the position of Asn residues at the subunit interface. The portions of the interface due to the Asn (green) and non-Asn (blue) residues of one subunit and the Asn 14, 65 and 78 residues (red) of the complementary subunit are depicted as dots of the same colors. (Redrawn from Ahern *et al.*, 1987.)

Asp residue, thereby promoting dissociation to monomers. Examination of the primary structure of triosephosphate isomerase from various organisms (Lu et al., 1984; Pichersky et al., 1984) revealed that small charged amino acid residues at these sites conserve the enzyme activity and that the thermophilic enzyme (Lu et al., 1984) does not have Asn these positions.

Ahern et al., (1987) selected three single mutations: Asn 78 to Ile, Thr of Asp (as a control) and a double mutation, Asn 14 to Thr and Asn 98 to Ile. With respect to irreversible thermal inactivation in the presence of guanidine hydrochloride, the single mutated protein with Ile or Thr 78 increased the half-life by 25%, and the half-life of double mutant increase about two times. The control mutant with Asp 87, on the other hand, decreased the activity relative to the wild-type protein.

Conformational stability with respect to reversible denaturation was then examined with the two single mutants, by the procedure as described above. As expected, the mutation Asn 78 to Thr or Ile had little effect on reversible denaturation, but the mutant with Asp 78 was destabilized.

6.3 MUTAGENESIS WITHOUT KNOWLEDGE OF PROTEIN STRUCTURE

In section 6.2 were described various procedures for stabilizing proteins of which X-ray crystal structures had been determined, with some typical examples. The substitution for amino acids in protein molecules is designed using computer modeling to avoid unnecessary conformational distortion in engineered proteins. The molecular reasons for the increased or decreased stability of engineered proteins and also for protein stability itself can be deduced by comparing the X-ray crystal structures of the wild-type and mutant proteins. The thermodynamic and kinetic analyses of the wild-type and mutant proteins will help in understanding the physico-chemical mechanism of protein stability through protein engineering.

Although the X-ray crystal structures of many proteins still remained undetermined, because of difficulty in crystallizing proteins, the amino acid sequence of a protein can be easily determined by sequencing the DNA encoding the protein. In such cases the stabilization of these proteins through site-directed mutagenesis can be possible, if a homologous protein of known tertiary structure is available, and the candidate substitution site is known.

6.3.1 Use of a homologous model protein

Neutral protease from B. stearothermophilus
Alteration of the thermostability of thermostable neutral protease from *B. stearothermophilus*, the sequence of which is known (Fujii et al., 1983; Takagi et al., 1985), was attempted by site-directed mutagenesis (Imanaka et al., 1986).

The principle for protein design is that substitutions made to increase internal hydrophobicity and to stabilize helices for strong internal packing enhance protein thermostability (Argos et al., 1979; Menendez-Arias and Argos, 1989). To select a model protein for designing to stabilize the protein, the amino acid sequences of four neutral proteases from *B. thermoproteolyticus* (thermolysin) (Titani et al., 1972), *B. subtilis* (Yang et al., 1984), *B. amyloliquefaciens* (Vasantha et al., 1984) and *B.*

6.3 Mutagenesis

stearothermophilus were compared. As shown in Fig. 6.20, the seqences of the two thermostable proteases from *B. thermoproteolyticus* and *B. stearothermophilus* exhibit high homology (85%), and the homology between the other two thermolabile enzymes is also high (89%). In contrast, the homology between the thermostable and thermolabile enzymes is low ($\sim 45\%$). The amino acid frames of the *B. stearothermophilus* enzyme and thermolysin were completely matched except for an additional three amino acids (Gly 29–Tyr 30–Tyr 31) in the former enzyme (Fig. 6.20). Therefore, the X-ray structure of *B. stearothermophilus* protease would be basically similar to that of thermolysin, which has already been determined (Fig. 6.21) (Matthews et al., 1972).

```
                                                          - .         -+ .        Ca
1 VAGASTVGVGRGVLGDQKYINTTYSSYYGYYYLQDNTR  GSGIFTYDGRNRT VLPGSLWTD 60
2 ITGTSTVGVGRGVLGDQKNINTTYSTYY    YLQDNTR GDGIFTYDAKYRT TLPGSLWAD
3 AAA T GSGTTLKGATVPLN- ISTEGGKYVLRDLSKPTGTQIITYDLQNRQSRLPGTLVSS
4 AAT T GTGTTLKGKTVSLN  ISSESGKYVLRDLSKPTGTQIITYDLQNREYNLPGTLVSS

   A(M2)
   |   S(M3)
   -C  |
1 GDNQFTASYDA AAVDAHY YAGVVYDYYKNVHGRLSYDGSNAA IRSTVHYG RGYNNAFWNG 120
2 ADNQFFASYDA PAVDAHY YAGVTYDYYKNVHNRLSYDGNNAA IRSSVHYS QGYNNAFWNG
3 TTKTFTSSSQR AAVDAHY NLGKVYDYFYSNFKRNSYDNKGSK IVSSVHYG TQYNNAAWTG
4 TTNQFTTSSQR AAVDAHY NLGKVYDYFYQKFNRNSYDNKGGK IVSSVHYGS RYNNAAWIG

                 A(M1)
                 |
            .Ca  -Zn Zn .                        Zn .       Ca
1 SQMVYGDGDGQ TFLPFSGGI DVVGHELTH AVTDYTAGLVYQNESGAINEAMSDIFGTLVE 180
2 SEMVYGDGDGQ TFIPLSGGI DVVAHELTH AVTDYTAGLIYQNESGAINEAISDIFGTLVE
3 DQMIYGDGDGS FFSPLSGSL DVTAHEMTH GVTQETANLIYENQPGALNESFSDVFG
4 DQMIYGDGDGS FFSPLSGSM DVTAHEMTH GVTQETANLNYENQPGALNESFSDVFG

    Ca.  Ca Ca     .+Ca+
1 FYANRNP DWEIGEDI YTPGVAGD ALRSMSDPAKY GDPDHYSK RYT GTQDNGGVHTNSGII 240
2 FTANKNP DWEIGEDV YTPGISGD SLRSMSDPAKY GDPDHYSK RYT GTQDNGGVQINSGII
3 YFNDTE DWDIGEDI T VSQP ALRSLSNPTKY NQPDNYANYRNLPNTDEG DYGGVHTNSGIP
4 YFNDTE DWDIGEDI T VSQP ALRSLSNPTKY GQPDNFKNYKNLPNTDAC DYGGVHTNSGIP

                                                         .+
1 NKAAY LLSQGGVHYGVSVNGIGRDKMGK IFYRALVYYLTPTS NFSQLRAACV QAAADLYG 300
2 NKAAY LISQGGTHYGVSVVGIGRDKLGK IFYRALTQYLTPTS NFSQLRAAAV QSAYDLYG
3 NKAAY NTITK    LGVSKSQQ IYYRALTTYLTPSS TFKDAKAAL IQSARDLYG
4 NKAAY NTTTK    IGVNKAEQ IYYRALTVYLTPSS TFKDAKAAL IQSARDLYG

1 ST SQEVNSVKQAFNAVGVY 319
2 ST SQEVASVKQAFDAVGVK
3 ST D  AAKVEAAWNAVGL
4 SQ D  AASVEAAWNAVGL
```

Fig. 6.20—Comparison of amino acid sequences of various extracellular neutral proteases. A blank indicates the absence of corresponding amino acid of this position. The proteases are from *B. stearothermophilus* (1) (Takagi et al., 1985), *B. thermoproteolytics* (2) (Titani et al., 1972), *B. subtilis* (3) (Yang et al., 1984), and *B. amyloliquefaciens* (4) (Vasantha et al., 1984) (4). Active and substrate-binding sites of thermolysin are indicated by ○ and □, respectively. Protein ligands for Zn and Ca ions for thermolysin are indicated above the sequence. Substitutions expected to enhance or deduce the thermostability of *B. stearothermophilus* protease in comparison with that of thermolysin are indicated above the sequence by + or −, respectively. Vertical arrows indicate amino acid substitutions in *B. stearothermophilus* protease. (Redrawn from Imanaka et al., 1986.)

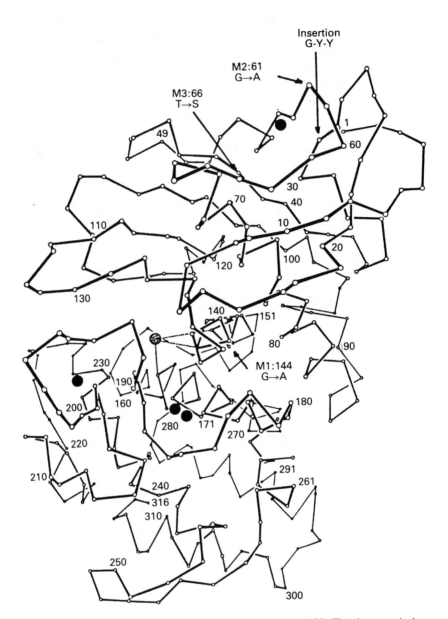

Fig. 6.21—Backbone structure of thermolysin (Matthews *et al.*, 1972). The zinc atom is drawn stippled with its three protein ligands shown diagrammatically as borken lines. Four calcium atoms are shown as solid circles. In *B. stearothermophilus* neutral protease three amino acids (Gly–Tyr–Tyr) are inserted between positions 28 and 29 in thermolysin. Amino acid substitutions in *B. stearothermophilus* protease are shown by arrows with the sequence number of substitution sites.

His 234 (active site), Arg 206 (substrate-binding site), Leu 205, Glu 193 and Asn 162 are near the active site in the three-dimensional struture of thermolysin and completely conserved in the four neutral proteases (Fig. 6.20). Therefore, these amino acids and others in the highly conserved sequences should not be altered.

The position and species of amino acids to be replaced to alter enzyme thermostability were then decided according to the preferable amino acid substitutions for stabilizing proteins (Argos *et al.*, 1979; Menendez-Arias and Argos, 1989). Thermolysin and *B. stearothermophilus* protease have some amino acids which promote higher thermostability in the first and second halves, respectively. The amino acid substitution in *B. stearothermophilus* protease was then confined to the first half. The possible substitution Gly to Ala is for positions 47, 61 and 144, and Gly 144 is located in the α-helix which combines two domains (Fig. 6.21). The replacement of Gly 144 by Ala (mutation M1) might increase the internal hydrophobicity, stabilize α-helix and thus enhance the thermostability of the protease. The substitution which adds only a methyl group at position 144 will minimize interruption of function or internal residue packing arrangement. Another mutation, M2 (Gly 61 to Ala), will

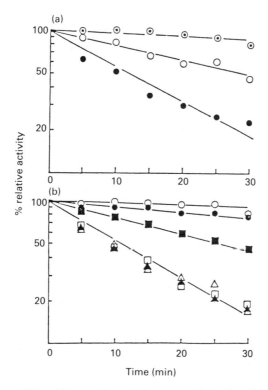

Fig. 6.22—Thermostability of *B. stearothermophilus* protease derivatives. Samples were heated at 75°C (a) or 65°C (b), and remaining activity was measured and exressed as the percentage of original activity. ⊙, Thermolysin; ●, wild-type of *B. stearothermophilus* protease; ○, mutant with Ala 144 (←Gly); △, mutant with Ser 66 (←Thr); ▲, double mutant with Ala 144 (←Gly) and Ser 66 (←Thr); □, double mutant with Ala 61 (←Gly) and Ser 66 (←Thr); ■, triple mutant with Ala (←Gly), Ser 66 (←Thr) and Ala 61 (←Gly). (Redrawn from Imanaka *et al.*, 1986.)

increase the thermostability, and mutation M3 (Thr 66 to Ser) should destabilize the protease (Argos et al., 1979; Menendez-Arias and Argos, 1989).

As predicted, the substitution Gly 144 to Ala (M1) gave additional stability to B. stearothermophilus protease (Fig. 6.22a), while the replacement of Thr 66 by Ser (M3) decreased the stability of the wild-type enzyme (Fig. 6.22b). The introduction of a stabilizing mutation (M1 and M2) to the destabilized mutant (M3) could not recover the thermostability of M3 (Fig. 6.22b). However, the thermostability of M3 was partly recovered by introducing both mutations M1 and M2 to M3. The results indicate that the effects on stability are not additive and may be cooperative.

Thermus aquaticus protease

T. aquaticus YT-1 produces a thermostable extracellular serine protease (aqualysin) (Matsuzawa et al., 1988). It was considered that one of the molecular reasons for higher stability of aqualysin than subtilisin is the two disulfide bonds between positions 67 and 99 and positions 163 and 194 in aqualysin (Matsuzawa et al., 1988; Takagi et al., 1990). Subtilisin has no disulfide bond (Markland and Smith, 1971). Rather high homology in amino acid sequence is observed between aqualysin and subtilisin (about 40%) (Takagi et al., 1990), although considerably lower than that between thermolysin and neutral protease from B. stearothermophilus (Imanaka et al., 1986). Introduction of a disulfide bond into subtilisin was then attempted using aqualysin as a model protein.

A pair of Gly 61 and Ser 98 was selected, because the distance between the two α-atoms is the most suitable for disulfide formation, which corresponds to the N-terminal disulfide bond between Cys 67 and Cys 99 in aqualysin.

The two Cys residues introduced by site-directed mutagenesis formed a disulfide bond in cells, and the oxidized mutant exhibited almost the same activity as that of wild-type protein. When the thermostability was measured in the temperature range 45–65°C, the half-life time was prolonged two or three times on cross-linking. The Cys-replaced mutant exhibited the same stability as that of wild-type protein. The T_m values of the cross-linked and wild-type enzymes are 63°C and 58.5°C, respectively. These results, although thermodynamic and kinetic analyses have not been made, may indicate that this is the first example of stabilizing subtilisin by cross-linkage.

6.3.2 Mutagenesis without a homologous protein

If an amino acid residue in protein has been correlated to its increased stability, site-directed mutagenesis can be applied to the protein, and the thermodynamic and/or kinetic analysis of the mutant proteins with different amino acid residues at the site will afford some insight into the mechanism of protein stability. The tertiary structures of E. coli tryptophan synthase α-subunit and kanamycin nucleotidyltransferase have not yet been determined. However, it has been shown that only a single amino acid substitution changes the stability of the proteins, which were deduced from correlating stability to amino acid sequence with the wild-type and naturally occurring variant proteins (tryptophan synthase α-subunit) (Yutani et al., 1977) or with mesophilic and thermophilic proteins (kanamycin nucleotidyltransferase) (Matsumura et al., 1984).

α-Subunit of tryptophan synthase

α-Subunits of tryptophan synthase from various mutants of *E. coli* differ in thermostability from the wild-type protein (Yutani *et al.*, 1977). The mutant protein in which Glu 49 (wild-type protein) is replaced with Met is more stable than the mutant protein toward urea-induced reversible denaturation. The difference in ΔG for protein unfolding was essentially the same as the difference in ΔG_{tra} for transferring from water to ethanol between the two amino acid residues. This is strongly indicative of the relation of the internal hydrophobicity to the stability of the protein. The position at 49 may be in the interior of the protein, as suggested from the X-ray crystal structure of tryptophan synthase $\alpha_2\beta_2$ complex of *Salmonella typhimurium* (Hyde *et al.*, 1988).

The substitution of various amino acids at position 49 was then attempted to examine the role of amino acid residue at the site in stabilizing the proteins. The mutant proteins with 20 substituted amino acids were constructed by site-directed mutagenesis (Yutani *et al.*, 1987). The stability of each resulting protein was calculated from the free energy changes between the native and unfolded proteins (thermodynamics) and between the native and transition states of unfolding (kinetics). Both measurements of stability correlated well with the dydrophobicity of the

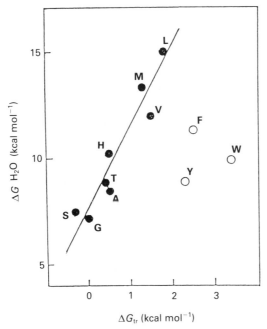

Fig. 6.23—Correlation between stability of wild-type and 18 mutants of tryptophan synthase α-subunit and hydrophobicity of the substituting residues at position 49. Amino acids designated by the single-letter code correspond to the data of proteins that have the indicated amino acids at position 49. Protein stability is represented by free energy change for protein unfolding (ΔG) at pH 7.0 and hydrophobicity of amino acid side chains is expressed by free energy change on transfer from water to ethanol (ΔG_{tr}). The straight line was obtained from least-squares fit of the eight points shown by solid circles. (Redrawn from Yutani *et al.*, 1987.)

substituted amino acids, except where the proteins contained introduced aromatic amino acid residues. Figure 6.23 shows the linear relationship between thermodynamic stability at pH 7.0 and hydrophobicity of substituted amino acid residues. The slope was 3.71. Similar linearity was observed at pH 9.0, but the slope was 1.9. It was concluded from the results that increasing the internal hydrophobicity *per se* stabilized the protein thermodynamically and kinetically. However, it was noticed that some extra energy which resulted from increased hydrophobicity might also contribute to increased stability, because the slopes were different at pH 7.0 and 9.0, and both slopes differed from unity.

The stabilities of the proteins with introduced aromatic amino acid residues were lower than would be predicted from consideration of the hydrophobicity of the residues (Fig. 6.23). This may be due to the conformational distortion around the bulky substitution site.

Kanamycin nucleotidyltransferase

Kanamycin nucleotidyltransferase is encoded by a bacterial plasmid pUB110 (Sadaie *et al.*, 1980). The comparison of amino acid sequences between thermophilic and mesophilic enzymes revealed that a single amino acid substitution—replacing Asp 80 with Tyr—is the cause of enhanced thermostability of the former relative to the latter (Matsumura *et al.*, 1984). To reveal the role of Tyr 80 in stabilizing the protein, the mutant proteins with Ser, Thr, Ala, Val, Leu, Phe and Trp at position 80 were constructed by site-directed mutagenesis (Matsumura *et al.*, 1988b).

These eight mutant proteins exhibited various half-life values for thermal inactivation at 58°C. The different half-life values correlated with the hydrophobicity of the substituted amino acid residues. The free energy of unfolding assessed from urea denaturation was also correlated with hydrophobicity. These results suggested that different amino acids at position 80 contribute to the stability of the protein by hydrophobic interactions. The stability of the mutant with Tyr 80 was unusually higher than that of the protein with Phe 80. This may be due to the contribution of the hydroxyl group of Tyr to the conformational stability.

Similarly to tryptophan synthase α-subunit (Yutani *et al.*, 1987), linearity between protein stability and hydrophobicity of substituted amino acid residues was observed with kanamycin nucleotidyltransferase, and the slope was more than unity. This also indicates the contribution of factors other than increased hydrophobic interaction resulting from amino acid substitution in kanamycin nucleotidyltransferase.

In contrast, an increase of internal hydrophobicity by amino acid substitution in T4 lysozyme was reported to be a major contribution to additional stabilization of the protein (Matsumura *et al.*, 1988a; see section 6.2.2.1).

Iso-1-cytochrome c

Iso-1- cytochrome c of *Saccharomyces cerevisiae* can be mutated by the application of ultraviolet (UV) irradiation or chemicals to the yeast (Hampsey *et al.*, 1986). Das *et al.* (1989) produced numerous mutants and their revertants from the yeast by the use of above mutagens. One of the mutant proteins (*cycl-73* mutant) in which natural Gly 34 is replaced with Ser was UV irradiated. Two kinds of mutation occurred;

one contained the reversion of Ser 34 to Gly, and another the conversion of Asn 57 to Ile with unchanged replacement, Gly 34 to Ser. Another mutant *cycl-190* contained a replacement of His 38 to Pro. Analysis of the revertants raised from the mutant by UV or chemical mutagens revealed the replacement, Asn 57 to Ile, together with the replacements at the original site (Pro 38 to Leu, Ser or Ala). The results suggested that the Ile 57 replacement appears to stabilize the Ser 34 and Pro 38 mutant proteins. An iso-1-cytochrome c having only an Asn 57 to Ile replacement was constructed by site-directed mutagenesis. During the DSC experiments to measure T_m, free Cys 107 caused dimerization of protein samples, which complicated the accurate determination of T_m. The free Cys 107 in the Asn 57 and Ile 57 proteins were then blocked with methyl methanethiosulfonate.

As shown in Table 6.13, the Ile 57-substituted mutant protein was more thermostable toward reversible thermal denaturation than the wild-type protein. The values of T_m and ΔG for the mutant were 17.2°C and 4.2 kcal mol^{-1} greater than the wild-type protein with Asn 57, respectively. Such a dramatic increase in stability through site-directed mutagenesis was rare (see many examples in this chapter). Two possible explanations for the enhanced stability of yeast can be considered: Asn 57 is a helix-breaking residue, while the replacement of Asn with Ile is a better helix former than Asn and is frequently observed in the middle of an α-helix (Chou and Fasman, 1978); and another is the greater hydrophobicity of Ile relative to Asn. The difference in free energy changes on transferring the Ile and Asn side chains from ethanol to water is 3 kcal mol^{-1}, which could account for approximately 70% of the increase in ΔG. The two reasons, however, seem not sufficient for the explanation of such dramatic thermostabilization by a single amino acid replacement. Further studies should be made on other amino acid replacements at position 57 and other sites, correlating to the changes of stability and X-ray crystal structure.

Table 6.13—Thermodynamic parameters for thermal unfolding of the normal with Asn 57 and mutant with Ile 57 of iso-1-cytochrome c. Cys 107 of both proteins were blocked with methyl methanethiosulfonate, and SCH$_3$-modified proteins were used (see text). The proteins were thermally denatured and the transitions were monitored by the absorbance change at 287 nm. The resultant absorbance versus temperature data were evaluated for van't Hoff enthalpy and entropy change for thermal unfolding at the midpoint temperature. T_m was also calculated. *ΔH and ΔS are the average values obtained by van't Hoff analysis of two unfolding experiments using independently prepared samples. $^\dagger \Delta G$ at 25°C was calculated by using the presented ΔH and ΔS values with change in ΔC_p, which was estimated from a plot of T_m versus ΔH (Redrawn from Das *et al.*, 1989)

Protein	T_m(°C)	ΔH (kcal mol^{-1})*	ΔS (cal mol^{-1} K^{-1})*	ΔG at 250°C (kcal mol^{-1})†
(Asn^{57})Cyt–SCH$_3$	46.5 ± 1.0	79 ± 5	247 ± 14	3.8
(Ile^{57})Cyt–SCH$_3$	63.7 ± 1.0	109 ± 9	323 ± 27	8.0

6.4 STABILIZATION THROUGH OTHER MUTAGENESIS

If neither the tertiary structure of a protein nor of any homologous model protein are available nor is there any hint for amino acid substitution, other strategies could be considered for stabilizing proteins. The strategies are (1) chemical mutagenesis of the gene (or the plasmid of phage harboring the gene) (Matsumura and Aiba, 1985; Matsumura *et al.*, 1986b) or organisms (Nishiyama *et al.*, 1986) and (2) thermal adaptation of the gene in microorganisms by exposing the organisms to higher temperatures (Hendrix and Welker, 1985) or thermal mutation of the gene transformed in a thermophile at higher temperature (Liao *et al.*, 1986).

6.4.1 Chemical mutagenesis

Kanamycin nucleotidyltransferase

A kanamycin nucleotidyltransferase is encoded by a plasmid pUB110, which was isolated from a mesophile (Lacey and Chopra, 1974). Another plasmid pTB913 encoding for a similar enzyme was isolated from a thermophile (Imanaka *et al.*, 1984; Matsumura *et al.*, 1984), and the enzyme encoded by pTB913 was shown to be more thermostable than that encoded by pUB110. The difference in stability between the two enzymes was attributed to a single amino acid replacement (Thr 130 in a mesophilic enzyme to Lys in a thermophilic enzyme) in their sequences (Matsumura *et al.*, 1984). The result indicates that the replacement of even a single amino acid in a mesophilic protein could increase its thermostability. The stabilization of the enzyme by site-directed mutagenesis was described previously.

The mesophilic gene, harbored on M13mp10 phage, was mutagenized with hydroxylamine (Matsumura and Aiba, 1985). In order to isolate the mutagenized gene which may encode a thermostable enzyme, the mutant gene was recloned in a vector plasmid, pBT922. The plasmid can multiply both in mesophile and thermophile conditions. The recombinant plasmid was then transformed into *B. stearothermophilus*. The clone containing the mutant gene encoding the more thermostable enzyme was isolated by shifting the growth temperature of the thermophile from 55 to 65°C. Such a temperature shift inactivates the wild-type enzyme (the mesophilic enzyme). Sequencing of the DNA of the mutant genes revealed that two types of mutation occurred on the original mesophilic gene, indicative of two types of amino acid replacement. The first type of mutation was Asp 80 to Tyr. Another mutation was Thr 130 to Lys. The Thr 130 to Lys mutation was the same replacement as observed on comparison of the mesophilic and thermophilic enzymes (Matsumura *et al.*, 1984). Both mutant enzymes were more thermostable than the wild-type, mesophilic protein.

6.4.2 Thermal mutagenesis

T4 lysozyme

The application of mutagens such as 5-bromouracil or 2-aminopurine to T4 lysozyme produced temperature-sensitive mutants (Alber *et al.*, 1987b). The X-ray crystal structures of the mutants showed that a single mutation occurs and that the amino acid substitutions are observed at residues which have low mobility, estimated from thermal factors of the side chain and main chain atoms and have reduced static solvent accessibility of the side chain in the folded state. The substitutions are varied in the properties of H-bonding, net charge, amino acid size and hydrophobic surface area. This indicates that there is no simple pattern in the nature of the amino acid substitution that causes temperature sensitivity. Similar observations were reported with cro phage λ protein (Pakula et al., 1986), staphylococcal nuclease (Shortle and Lin, 1985), and kanamycin nucleotidyltransferase (Matsumura et al., 1986a). Many different types of non-covalent interactions such as ion pairs, H-bonds, van der Waals contacts and hydrophobic contacts can make quantitatively comparable contributions to the stability of a protein.

Mobile or exposed amino acids seem to be more tolerant of mutations in T4 lysozyme. Ala 96 and Thr 157 in the emzyme are the most mobile and exposed sites of temperature-sensitive mutation. The substitutions at the positions 96 (Alber et al., 1987b) and 157 (Grutter et al., 1979; Alber and Matthews, 1987), however, produced temperature-sensitive mutants. These suggested that the most critical substitutions are the methylene chain or guanidinium nitrogen of Arg 96 and γ-hydroxyl group of Thr 157. These atoms are among the least mobile or least accessible in their respective side chain.

Crystallographic thermal factors and calculated solvent accessibility are highly correlated, so it is difficult to determine if one of the parameters is more important than the other (Westhoff et al., 1984; Tainer et al., 1984; Sheriff et al., 1985).

Kanamycin nucleotidyltransferase

When *B. stearothermophilus* was exposed to elevated temperature, a restriction endonuclease in the thermophile became more thermostable, due to the temperature-induced mutagenesis of the enzyme gene (Hendrix and Welker, 1985). Such biological adaptation may be effective to produce thermostable protein *in vivo*, without any knowledge of the information on protein structure.

Kanamycin nucleotidyltransferase gene was mutagenized using such a strategy (Liao et al., 1986). The plasmid, pUB110, carries the mesophilic kanamycin nucleotidyltransferase gene, as described above. Because of the thermolability of replication origin on the plasmid, a chimeric plasmid (pBST2) was constructed from pUB110 and pBST1 carrying a heat-stable origin of replication. pBST2 can therefore multiply in a thermophile at elevated temperature. The plasmid, pBST2, was transformed into *B. stearothermophilus*, and the thermophile was cultured at 63°C, at which temperature the kanamycin nucleotidyltransferase produced by the mesophile gene was inactivated. During the culture a thermally induced mutagenesis of the gene occurred, and Asp 80 in the native-type protein was replaced with Tyr—a single point mutation of GAT to TAT. When the growth temperature of the thermophile containing the mutant gene was elevated to 69–71°C, another type of mutation (Thr 130 to Lys) occurred, and a double mutated protein (Asp 80 to Tyr and Thr 130 to Lys) was produced. The thermostability of the double mutant enzyme was higher than that of the single mutant protein. The mutation sites in the enzyme introduced by *in vivo* mutagenesis are the same as those found in the variant enzymes produced by *in vitro* mutation (Matsumura and Aiba, 1985). In addition, one of the mutations is the naturally occurring one appearing in the thermophile (Imanaka et al., 1984; Matsumura et al., 1984).

REFERENCES

Ahern, T. J. and Klibanov, A. K. (1985) The mechanism of irreversible enzyme inactivation at 100°C. *Science*, **228**, 1280–1284.

Ahern, T. J., Casal, J. I., Petsko, G. A. and Klibanov, A. M. (1987) Control of oligomeric enzyme thermostability by protein engineering. *Proc. Natl. Acad. Sci. USA*, **84**, 675–679.

Alber, T. and Matthews, B. W. (1987) Structure and thermal stability of phage T4 lysozyme. *Methods Enzymol.*, **154**, 511–533.

Alber, T., Banner, D. W., Bloomer, A. C., Petsko, G. A., Philip, R. S. D., Rivers, P. S. and Wilson, I. A. (1981) On the three-dimensional structure and catalytic mechanism of triose phosphate isomerase. *Phil. Trans. R. Soc. London B*, **293**, 159–171.

Alber, T., Dao-pin, S., Wilson, K., Wozniak, J. A., Cook, S. P. and Matthews, B. W. (1987a) Contributions of hydrogen bonds of Thr 157 to the thermodynamic stability of phage T4 lysozyme. *Nature (London)*, **330**, 41–46.

Alber, T., Dao-pin, S., Nye, J. A., Muchmore, D. C. and Matthews, B. W. (1987b) Temperature-sensitive mutations of bacteriophage T4 lysozyme occur at sites with low mobility and low solvent accessibility in the folded protein. *Biochemistry*, **26**, 3754–3758.

Argos, P., Rossman, M. G., Grau, U. M., Zuber, H., Frank, G. and Tratschin, J. D. (1979) Thermal stability and protein structure. *Biochemistry*, **18**, 5698–5703.

Baldwin, R. L. (1986) Temperature dependence of the hydrophobic interaction in protein folding. *Proc. Natl. Acad. Sci. USA*, **83**, 8069–8072.

Becktel, W. J. and Baase, W. A. (1987) Thermal denaturation of bacteriophage T4 lysozyme at neutral pH. *Biopolymers*, **26**, 619–623.

Bernstein, F. C., Koetzle, T. F., Williams, G. J. B., Meger, E. F., Jr, Brice, M. D., Rodgers, J. R., Kennard, O., Shimanouchi, T. and Tasumi, M. (1977). The protein data bank: a computer-based archival file for macromolecular structures. *J. Mol. Biol.*, **112**, 535–542.

Bott, P.R., Ultsch, M., Kossiakoff, A., Graycar, T., Katz, B. and Power, S. (1988) The three-dimensional structure of *Bacillus amyloliquefaciens* subtilisin at 1.8Å and analysis of the structural consequences of peroxide inactivation. *J. Biol. Chem.*, **263**, 7895–7906.

Brot, N. and Weissbach, H. (1983) Biochemistry and physiological role of methionine sulfoxide residues in proteins. *Arch. Biochem. Biophys.*, **223**, 271–281.

Carter, P. and Wells, J. A. (1988) Dissecting the catalytic triad of a serine protease. *Nature (London)*, **332**, 564–568.

Chothia, C. (1974) Hydrophobic bonding and accessible surface area in proteins. *Nature (London)*, **248**, 338–339.

Chou, P. Y. and Fasman, G. D. (1978) Prediction of the secondary structure of proteins from their amino acid sequence. *Adv. Enzymol.*, **47**, 45–148.

Colman, P. M., Jansonius, K. N. and Matthews, B. W. (1972) Structure of thermolysin: electron density map at 2.3 Å resolution. *J. Mol. Biol.*, **70**, 701–724.

Craik, C. S., Largman, C., Fletcher, T., Roczniak, S., Barr, P. J., Fletterick, R. and Rutter, W. J. (1985) Redesigning trypsin: alteration of substrate specificity. *Science*, **228**, 291–297.

Creighton, T. E. (1983) *Proteins*. Freeman, New York.

Das, G., Hickey, D. R., McLendon, D., McLendon, G. and Sherman, F. (1989) Dramatic thermostabilization of yeast iso-1-cytothrome c by an asparagine → isoleucine replacement at position 57. *Proc. Natl. Acad. Sci. USA*, **86**, 496–499.

Drenth, J., Hol, W. G. J. and Jansonius, J. (1972) Subtilisin novo: three-dimensional structure and its comparison with subtilisin BPN'. *Eur. J. Biochem.*, **26**, 177–181.

Dillon, J. R., Nasim, A. and Nestmann, E. R. (eds) (1985) *Recombinant DNA methodology*. Wiley, New York.

Eisenberg, D. and McLachlan, A. D. (1986) Covalent energy in protein folding and unfolding. *Nature (London)*, **319**, 199–203.

Erwin, C. R., Barnett, B. L., Oliver, J. D. and Sullivan, J. F. (1990) Effects of engineered salt bridges on the stability of subtilisin BPN'. *Protein Eng.*, **4**, 87–97.

Estell, D. A., Graycar, T. P. and Wells, J. A. (1985) Engineering an enzyme by site-directed mutagenesis to be resistant to chemical oxidation. *J. Biol. Chem.*, **260**, 6518–6521.

Fersht, A. R., Shi, J-P, Knill-Jones, J., Lowe, D. M., Wilkinson, A. J., Blow, D. M., Brick, P., Caarter, P., Waye, M. M. Y. and Winter, G. (1985) Hydrogen bonding and biological specificity analysed by protein engineering. *Nature (London)*, **314**, 235–238.

Fujii, M., Takagi, M., Imanaka, T. and Aiba, S. (1983) Molecular cloning of a thermostable neutral protease gene from *Bacillus stearothermophilus* in a vector plasmid and its expression in *Bacillus stearothermophilus* and *Bacillus subtilis*. *J. Bacteriol.*, **154**, 831–837.

Ghosh, S. S., Bocks, S. C., Rokita, S. E. and Kaiser, E. T. (1986) Modification of the active site of alkaline phosphatase by site-directed mutagenesis. *Science*, **231**, 145–148.

Goto, Y. and Hamaguchi, K. (1979) The role of the intrachain disulfide bond in the conformation and stability of the constant fragment of the immunoglobulin light chain. *J. Biochem. (Tokyo)*, **86**, 1433–1441.

Grutter, M. G., Hawkes, R. B. and Matthews, B. W. (1979) Molecular basis of thermostability in the lysozyme from bacteriophage T4. *Nature (London)*, **277**, 667–669.

Grutter, M. G., Gray, T. M., Weaver, L. H., Alber, T., Wilson, K. and Matthews, B. W. (1987) Structural studies of mutants of the lysozyme of bacteriophage T4: the temperature-sensitive mutant protein Thr 157 to Ile. *J. Mol. Biol.*, **197**, 315–329.

Hampsey, D. M., Das, G. and Sherman, F. (1986) Amino acid replacements in yeast iso-1-cytochrome c: comparison with the phylogenetic series and the tertiary structure of related cytochromes. *J. Biol. Chem.*, **261**, 3259–3271.

Hanna, M. H. (1987) Applied genetics for biochemical engineering: recombinant DNA. In *Advanced biochemical engineering* (H. R. Bungay and G. Belfort, eds), pp. 103—128. Wiley, New York.

Hawkes, R., Grutter, M. G. and Schellman, J. (1984) Thermodynamic stability and point mutations of bacteriophage T4 lysozyme. *J. Mol. Biol.*, **175**, 195–212.

Hecht, M. H., Sturtevent, J. M. and Sauer, R. T. (1984) Effect of single amino acid replacements on the thermal stability of the NH_2-terminal domain of phage λ repressor. *Proc. Natl. Acad. Sci. USA*, **81**, 5685–5689.

Hecht, M. H., Hehir, K. M., Nelson, H. C. M., Sturtevant, J. M. and Sauer, R. T. (1985) Increasing and decreasing protein stability: effects of revertant substitutions on the thermal denaturation of phage λ repressor. *J. Cell. Biochem.*, **29**, 217–224.

Hecht, M. H., Sturtevant, J. M. and Sauer, R. T. (1986) Stabilization of λ repressor against thermal denaturation by site-directed Gly → Ala changes in α-helix 3. *Proteins Struct. Funct. Genet.*, **1**, 43–46.

Hendrix, J. D. and Welker, N. E. (1985) Isolation of a *Bacillus stearothermophilus* mutant exhibiting increased thermostability in its restriction endonuclease. *J. Bacteriol.*, **162**, 682–692.

Hol, W. G. J., van Duijinen, P. T. and Berendsen, H. J. C. (1978) The α-helix dipole and the properties of proteins. *Nature (London)*, **273**, 443–446.

Hyde, C. C., Ahmed, S. A., Padlan, E. A., Miles, E. W. and Davies, D. R. (1988) Three-dimensional structure of the tryptophan synthase alpha 2 beta 2 multienzyme complex from *Salmonela typhimurium*. *J. Biol. Chem.*, **263**, 17857–17871.

Imanaka, T., Ano, T., Fujii, M. and Aiba, S. (1984) Two replication determinants of an antibiotic-resistance plasmid, pTB19. *J. Gen. Microbiol.*, **130**, 1399–1408.

Imanaka, T., Shibazaki, M. and Takagi, M. (1986) A new way of enhancing the thermostability of proteases. *Nature (London)*, **324**, 695–697.

Johnson, R. E., Adams, P. and Rupley, J. A. (1978) Thermodynamics of protein cross-link. *Biochemistry*, **17**, 1479–1484.

Katz, B. A. and Kossiakoff, A. A. (1986) The crystallographically determined structure of a typical strained disulfide engineered into subtilisin. *J. Biol. Chem.*, **261**, 15480–15485.

Kellis, J. T., Nyberg, K., Jr, Sali D. and Fersht, A. R. (1988) Contribution of hydrophobic interactions to protein stability. *Nature (London)*, **333**, 784–786.

Klibanov, A. M. (1983) Stabilization of enzyme against thermal inactivation. *Adv. Appl. Microbiol.*, **29**, 1–28.

Kunkel, T. A. (1985) Rapid and efficient site-specific mutagenesis without phenotypic selection. *Proc. Natl. Acad. Sci. USA*, **82**, 488–492.

Kunkel, T. A., Roberts, J. D. and Zakour, R. A. (1987) Rapid and efficient site-specific mutagenesis without phenotypic selection. *Methods Enzymol.*, **154**, 367–382.

Lacey, R. W. and Chopra, J. (1974) Genetic studies of a multiresistant strain of a *Staphylococcus aureus*. *J. Med. Microbiol.*, **7**, 285–297.

Lee, B. and Richards, F. M. (1971) The interpretation of protein structure: estimation of static accessibility. *J. Mol. Biol.*, **55**, 379–400.

Levitt, M. (1983) Molecular dynamics of native proteins. I. Computer simulation of trajectories. *J. Mol. Biol.*, **168**, 595–620.

Liao, H., McKenzie, T. and Hageman, R. (1986) Isolation of a thermostable enzyme variant by cloning and selection in a thermophile. *Proc. Natl. Acad. Sci. USA*, **83**, 576–580.

Lin, S. H., Konishi, Y., Denton, M. E. and Scheraga, H. A. (1984) Influence of an extrinsic crosslink on the folding pathway of ribonuclease A: conformational and thermodynamic analysis of cross-linked (lysine 7, lysine 41)-ribonuclease A. *Biochemistry*, **23**, 5504–5512.

Lu, H. S., Yuan, M. P. and Gracy, R. W. (1984) Primary structure of human triosephosphate isomerase. *J. Biol. Chem.*, **259**, 11958–11968.

Lumry, R. and Rajender, S. (1970) Enthalpy–entropy compensation phenomena in water solutions of proteins and small molecules: a ubiquitous property of water. *Biopolymers*, **9**, 1125–1227.

Markland, F. S. and Smith, E. L. (1971) Subtilisins: primary structure, chemical and physical properties. *Enzymes*, **3**, 561–608.

Matsumura, M. and Aiba, S. (1985) Screening for thermostable mutant of kanamycin nucleotidyltransferase by the use of transformation system for a thermophile, *Bacillus stearothermophilus*. *J. Biol. Chem.*, **260**, 15298–15303.

Matsumura, M., Katakura, Y., Imanaka, T. and Aiba, S. (1984) Enzymatic and nucleotide sequence studies of a kanamycin-inactivating enzyme encoded by a plasmid from thermophilic bacilli in comparison with that encoded by plasmid pUB110. *J. Bacteriol.*, **160**, 413–420.

Matusmura, M., Kataoka, S. and Aiba, S. (1986a) Single amino acid replacements affecting the thermostability of kanamycin nucleotidyltransferase. *Mol. Gen. Genet.*, **204**, 355–358.

Matsumura, M., Yasumura, S. and Aiba, S. (1986b) Cumulative effect of intragenic amino acid replacements on the thermostability of a protein. *Nature (London)*, **323**, 356–358.

Matsumura, M., Becktel, W. and Matthews, B. W. (1988a) Hydrophobic stabilization in T4 lysozyme determined directly by multiple substitutions of Ile 3. *Nature (London)*, **334**, 406–410.

Matsumura, M., Yahanda, S., Yasumura, S., Yutani, K. and Aiba, S. (1988b) Role of tyrosine-80 in the stability of kanamycin nucleotidyltransferase analyzed by site-directed mutagenesis. *Eur. J. Biochem.*, **171**, 715–720.

Matsumura, M., Becktel, W. J., Levitt, M. and Matthews, B. W. (1989) Stabilization of phages T4 lysozyme by engineered disulfide bonds. *Proc. Natl. Acad. Sci. USA*, **86**, 6562–6566.

Matsuzawa, H., Tokugawa, K., Hamaoki, M., Mizoguchi, M., Tagauchi, H., Terada, I., Known S. T. and Ohta, T. (1988) Purification and characterization of aqualysin I (a thermophilic alkaline serine protease) produced by *Thermus aquaticus* YT-1. *Eur. J. Biochem.*, **171**, 441–447.

Matthews, B. W. (1987) Genetic and structural analysis of the protein stability problem. *Biochemistry*, **26**, 6885–6888.

Matthews, B. W. and Remington, S. J. (1974) The three dimensional structure of the lysozyme from bacteriophage T4. *Proc. Natl. Acad. Sci. USA*, **71**, 4178–4182.

Matthews, B. W., Jansonius, J. N., Coleman, P. M., Shoenborn, B. P. and Dupourque, D. (1972) Three-dimensional structure of thermolysin. *Nature New Biol. (London)*, **238**, 37–41.

Matthews, B. W., Remington, S. J., Gruetter, W. F. and Anderson, W. F. (1981) Relation between hen egg white lysozyme and bacteriophage T4 lysozyme. evolutionary implications. *J. Mol. Biol.*, **147**, 545–558.

Matthews, B. W., Nicholson, H. and Becktel, W. J. (1987) Enhanced protein thermostability from site-directed mutations that decrease the entropy of unfolding. *Proc. Natl. Acad. Sci. USA*, **84**, 6663–6667.

Mauguen, Y., Hartley, R. Q., Dodson, E. J., Bricogne, G. G., Chothia, C. and Jack, A. (1982) Molecular structure of a new family of ribonuclease. *Nature (London)*, **297**, 162–164.

Meloun, B., Baudys, M., Kostka, V., Hausdorf, G., Frommel, C. and Hoehne, W. E. (1985) Complete primary structure of thermitase from *Thermoactinomyces vulgaris* and its structural features related to the subtilisin-type proteinase. *Nature (London)*, **297**, 162–164.

Menendez-Arias, L. and Argos, P. (1989) Engineering protein thermal stability: sequence statistics point to residue substitutions in α-helices. *J. Mol. Biol.*, **206**, 397–406.

Miller, J. H. (1972) *Experiments in molecular genetics.* Cold Spring Harbor Laboratory, New York.

Mitchinson, C. and Baldwin, R. L. (1986) The design and production of semisynthetic ribonucleases with increased thermostability by incorporation of S-peptide analogues with enhanced helical stability. *Proteins Struct. Funct. Genet.*, **1**, 23–33.

Mitchison, C. and Wells, J. A. (1989) Protein engineering of disufide bonds in subtilisin BPN'. *Biochemistry*, **28**, 4807–4815.

Nemethy, G., Leach, S. J. and Scheraga, H. A. (1966) The influence of amino acid side chains on the free energy of helix-coil transition. *J. Phys. Chem.*, **70**, 998–1004.

Nicholson, H., Becktel, W. J. and Matthews, B. W. (1988) Enhanced protein thermostability from designed mutations that interact with α-helic dipoles. *Nature (London)*, **336**, 651–656.

Nishiyama, M., Matsubara, N., Yamamoto, K., Iijima, S., Uozumi, T. and Beppu, T. (1986) Nucleotide sequence of the malate dehydrogenase gene of *Thermus flavus* and its mutation directing an increase in enzyme activity. *J. Biol. Chem.*, **261**, 14178–14183.

Nosoh, Y. and Sekiguchi, T. (1988) Protein thermostability: mechanism and control through protein engineering. *Biocatalysis*, **1**, 257–273.

Pabo, C. O. and Lewis, M. (1982) The operator-binding domain of λ repressor: structure and DNA recognition. *Nature (London)*, **298**, 443–447.

Pace, C. N. (1986) Determination and analysis of urea and guanidine hydrochloride denaturation curve. *Meth. Enzymol.*, **131**, 266–279.

Paehler, A., Banerjie, A., Dattagupta, J. K., Fugita, T., Lindner, K., Pal, G. P., Suck, D., Weber, G. and Saenger, W. (1984) Three-dimensional structure of fungal proteinase K reveals similarity to bacterial subtilisin. *EMBO J.*, **3**, 1311–1314.

Pakula, A. A., Young, V. B. and Sauer, R. T. (1986) Bacteriophage *cro* mutations: effects on activity and intracellular degradation. *Proc. Natl. Acad. Sci. USA*, **83**, 8829–8833.

Pantoliano, M. W., Ladner, R. C., Bryan, P. N., Rollence, M. L., Wood, L. R. and Poulos, T. L. (1987) Protein engineering of subtilisin BPN': enhanced stabilization through the introduction of two cysteines to form a disulfide bond. *Biochemistry*, **26**, 2077–2082.

Perry, L. J. and Wetzel, R. (1984) Disulfide bond engineered into T4 lysozyme: stabilization of the protein toward thermal inactivation. *Science*, **226**, 555–557.

Perry, L J. and Wetzel, R. (1986) Unpaired cysteine-54 interferes with the ability of an engineered disulfide to stabilize T4 lysozyme. *Biochemistry*, **25**, 733–739.

Perry, L. J. and Wetzel, R. (1987) The role of cysteine oxidation in the thermal inactivation of T4 phage lysozyme. *Protein Eng.*, **1**, 101–105.

Perutz, M. F. (1978) Electrostatic effects in proteins. *Science*, **201**, 1187–1191.

Perutz, M. F. and Raidt, H. (1975) Stereochemical basis of heat stability in bacterial ferredoxins and in haemoglobin A2. *Nature (London)*, **255**, 256–259.

Pichersky, E., Gottlieb, L. D. and Hess, J. F. (1984) Nucleotide sequence of the triose phosphate isomerase gene of *Escherichia coli*. *Mol. Gen. Genet.*, **195**, 314–319.

Pjura, P. E., Matsumura, M., Wozniak, J. A. and Matthews, B. W. (1990) Structure of a thermostable disulfide-bridge mutant of phage T4 lysozyme shows that an engineered cross-link in a flexible region does not increase the rigidity of the folded protein. *Biochemistry*, **29**, 2592–2598.

Poland, D. C. and Scheraga, H. A. (1965a) Statistical mechanics of noncovalent bonds in polyamino acids. VIII. Covalent loops in protein. *Biopolymers*, **3**, 379–399.

Poland, D. C. and Scheraga, H. A. (1965b) Statistical mechanics of noncovalent bonds in polyamino acids. IX. The two-state theory of protein denaturation. *Biopolymers*, **3**, 401–419.

Privalov, P. L. (1979) Stability of proteins. *Adv. Protein Chem.*, **33**, 167–241.

Remington, S. J., Anderson, W. F., Owen, J., Grainger, C., Eyck, L. F. T. and Matthews, B. W. (1978) Structure of the lysozyme from bacteriophage T4: an electron density map at 2.4 Å resolution. *J. Mol. Biol.*, **118**, 81–98.

Richardson, J. S. (1981) The anatomy and taxonomy of protein structure. *Adv. Protein Chem.*, **34**, 167–339.

Richardson, J. S. and Richardson, D. C. (1988) Amino acid preferences for specific location at the end of α-helices. *Science*, **240**, 1648–1652.

Rose, G. D., Geselowitz, A. R., Lesser, G. J., Lee, R. H. and Zehfus, M. H. (1985) Hydrophobicity of amino acid residues in globular proteins. *Science*, **229**, 834–838.

Rossmann, M. G. and Argos, P. (1976) Exploring structure homology of proteins. *J. Mol. Biol.*, **105**, 75–95.

Ruegg, C., Ammer, D. and Lerch, K. (1982) Comparison of amino acid sequence and thermostability of tyrosinase from three wild type strains of *Neurospora crassa*. *J. Biol. Chem.*, **257**, 6420–6426.

Sadaie, Y., Burtis, K. C. and Doi, R. H. (1980) Purification and characterization of a kanamycin nucleotidyltransferase from pUB110-carrying cells of *Bacillus subtilis*. *J. Bacteriol.*, **141**, 1178–1182.

Schellman, J. A. (1978) Solvent denaturation. *Biopolymers*, **17**, 1305–1322.

Schulz, G. E. and Schirmer, R. H. (1979) *Principles of protein structure*. Springer, New York.

Sheriff, S., Hendrickson, W. A., Stenkamp, R. E., Siecker, L. C., and Jensen, L. H. (1985) Influence of solvent accessibility and intermolecular contacts on atomic mobilities in hemerythrins. *Proc. Natl. Acad. Sci. USA*, **82**, 1104–1107.

Shortle, D. (1986) Guanidine hydrochloride denaturation studies of mutant forms of staphylococcal nuclease. *J. Cell. Biochem.*, **30**, 281–289.

Shortle, D. (1989) Probing the determinants of protein folding and stability with amino acid substitutions. *J. Biol. Chem.*, **264**, 5315–5318.

Shortle, D. and Lin, B. (1985) Genetic analysis of staphylococcal nuclease: identification of three intragenic 'global' suppressors of nuclease-minus mutations. *Genetics*, **110**, 539–555.

Shortle, D. and Meeker, A. K. (1986) Mutant forms of staphylococcal nuclease with altered patterns of guanidine hydrochloride and urea denaturation. *Proteins Struct., Funct., Genet.*, **1**, 81–89.

Shortle, D., Meeker, A. K. and Freire, E. (1988) Stability mutants of staphylococcal nuclease: large compensating enthalpy–entropy changes for the reversible denaturation reaction. *Biochemistry*, **27**, 4761–4768.

Stauffer, C. E. and Eston, D. (1969) The effect on subtilisin activity of oxidizing a methionine residue. *J. Biol. Chem.*, **244**, 5333–5338.

Tainer, J. A., Getzoff, E. D., Alexander, H., Houghten, R. A., Olson, A. J., Lerner, R. A. and Hendrickson, W. A. (1984) The reactivity of anti-peptide antibodies is a function of the atomic mobility of sites in a protein. *Nature (London)*, **312**, 127–133.

Takagi, K., Imanaka, T. and Aiba, S. (1985) Nucleotide sequence and promoter region for the neutral protease gene from *Bacillus stearothermophilus*. *J. Bacteriol.*, **163**, 824–831.

Takagi, H., Takahashi, T., Momose, H., Inouye, M., Maeda, Y., Matsuzawa, H. and Ohta, T. (1990) Enhancement of subtilisin E by introduction of disulfide bond engineered on the basis of structural comparison with a thermophilic serine protease. *J. Biol. Chem.*, **265**, 6874–6678.

Takahashi, K. and Sturtevant, J. M. (1981) Thermal denaturation of streptomyces subtilisin inhibitor, subtilisin BPN', and the inhibitor-subtilisin complex. *Biochemistry*, **20**, 6185–6190.

Tanford, C. (1962) Contribution of hydrophobic interactions to the stability of the globular conformation of proteins. *J. Am. Chem. Soc.*, **84**, 4240–4247.

Thronton, J. M. (1981) Disulfide bridges in globular proteins. *J. Mol. Biol.*, **151**, 261–287.

Titani, K., Hermodson, M. A., Ericsson, L. H., Walsh, K. A. and Neurath, H. (1972) Amino acid sequence of thermolysin. *Nature New Biol. (London)*, **238**, 35–37.

Ueda, T., Yamada, H., Hirata, M. and Imoto, T. (1985) An intramolecular cross-linkage of lysozyme: formation of cross-links between lysine-1 and histidine-15 with bis(bromoacetamide) derivatives by a two-state reaction procedure and properties of the resulting derivatives. *Biochemistry*, **24**, 6316–6322.

Ulmer, K. M. (1983) Protein engineering. *Science*, **219**, 666–671.

Vasantha, N., Thompson, L. D., Rhodes, C., Banner, C., Nagle, J. and Filpula, D. (1984) Genes for alkaline protease and neutral protease from *Bacillus amyloliquefaciens* contain a large open reading frame between the regions coding for signal sequence and mature protein. *J. Bacteriol.*, **159**, 811–819.

Villafranca, J. E., Howell, E. E., Oately, S. J., Xuong, N.-H. and Kraut, J. (1987) An engineered disulfide bond in dihydrofolate reductase. *Biochemistry*, **26**, 2182–2189.

Voordouw, G., Milo, C. and Roche, R. S. (1976) Role of bound calcium ions in thermostable, proteolytic enzymes: separation of intrinsic and calcium ion contributions to the kinetic thermal stability. *Biochemistry*, **15**, 3716–3724.

Walker, J. E., Wonacott, A. J. and Harris, J. I. (1980) Heat stability of a tetrameric enzyme: D-glyceraldehyde-3-phosphate dehydrogenase. *Eur. J. Biochem.*, **108**, 581–586.

Wang, A., Lu, S. D. and Mark, D. F. (1984) Site-directed mutagenesis of human interleukin-2-gene: structure–function analysis of the cysteine residues. *Science*, **224**, 1431–1433.

Weaver, L. H. and Matthews, B. W. (1987) Structure of bacteriophage T4 lysozyme refined at 1.7 Å resolution. *J. Mol. Biol.*, **193**, 189–199.

Weiner, S. J., Kollman, P. A., Case, D. A., Singh, U. C., Ghio, C., Alagona, G., Profeta, S., Jr and Weiner, P. (1984) A new force field for molecular mechanical simulation of nucleic acid and proteins. *J. Am. Chem. Soc.*, **106**, 765–784.

Wells, T. N. C. and Fersht, A. R. (1985) Hydrogen bonding in enzymatic catalysis analyzed by protein engineering. *Nature (London)*, **316**, 656–657.

Wells J. A. and Powers, D. B. (1986) *In vivo* formation and stability of engineered disulfide bonds in subtilisin. *J. Biol. Chem.*, **261**, 6564–6570.

Wells, J. A., Ferrari, E., Henner, D. J., Estell, D. A. and Chen, F. Y. (1983) Cloning, seqencing, and secretion of *Bacillus amyloliquefaciens* subtilisin in *Bacillus subtilis*. *Nucleic Acid Res.*, **11**, 7911–7925.

Wells, J. A., Powers, D. B., Bott, R. R., Graycar, T. P. and Estell, D. A. (1987) Designing substrate sepcificity by protein engineering of electrostatic interactions. *Proc. Natl. Acad. Sci. USA*, **84**, 1219–1223.

Westhoff, E., Altschuh, D., Moras, D., Bloomer, A. C., Mondragon, A., Klug, A. and van Regenmortel, M. H. V. (1984) Correlation between segmental mobility and the locale of antigenic determinants in proteins. *Nature (London)*, **311**, 123–126.

Wetzel, R. (1987) Harnessing disulfide bonds using protein engineering. *Trends Biochem. Sci.*, **12**, 478–482.

Wetzel, R., Perry, L. J., Baase, W. A. and Becktel, W. J. (1988) Disulfide bonds and thermal stability in T4 lysozyme. *Proc. Natl. Acad. Sci. USA*, **85**, 401–405.

Wilkinson, A. J., Fersht, A. R., Blow, D. M. and Winter, G. (1983) Site-directed mutagenesis as a probe of enzyme structure and catalysis: tyrosyl-tRNA synthetase cysteine-35 to glycine-35 mutation. *Biochemistry*, **22**, 3581–3586.

Wilkinson, A. J., Fersht, A. R., Blow, D. M., Carter, P. and Winter, G. A. (1984) A large increase in enzyme–substrate affinity by protein engineering. *Nature (London)*, **307**, 187–188.

Wilks, H. M., Hart, K. W., Feeney, R., Dunn, C. R., Muirhead, H., Chia, W. N., Barstow, D. A., Atkinson, T., Clarke, A. R. and Holbrook, J. J. (1988) A specific, highly active malate dehydrogenase by redesign of a lactate dehydrogenase framework. *Science*, **242**, 1541–1544.

Winter, G., Fersht, A. R., Wilkinson, A. J., Zoller, M. and Smith, M. (1982) Redesigning enzyme structure by site-directed mutagenesis: tyrosyl tRNA synthetase and ATP binding. *Nature (London)*, **299**, 756–758.

Wright, C. S., Alden, R. A. and Kraut, J. (1969) Structure of subtilisin BPN' at 2.5-angstrom resolution. *Nature (London)*, **221**, 235–242.

Yang, M. Y., Ferrari, E. and Henner, D. J. (1984) Cloning of the neutral protease gene of *Bacillus subtilis* and the use of the cloned gene to create an *in vitro*-derived deletion mutation. *J. Bacteriol.*, **160**, 15–21.

Yutani, K., Ogasawara, K., Sugino, Y. and Matsushiro, A. (1977) Effect of a single amino acid substitution on stability of conformation of a protein. *Nature (London)*, **267**, 274–275.

Yutani, K., Ogasawara, K., Tsujita, T. and Sugino, Y. (1987) Dependence of conformational stability on hydrophobicity of the amino acid residue in a series of variant proteins substituted at a unique position of tryptophan synthetase α subunit. *Proc. Natl. Acad. Sci. USA*, **84**, 4441–4444.

Zale, D. E. and Klibanov, A. M. (1986) Why does ribonuclease irreversibly inactivate at high temperature? *Biochemistry*, **25**, 5432–5444.

7
Concluding remarks

The stability and stabilization of proteins have been and are still now attractive topics in the fundamental and applied fields. In this book, then, how proteins are stabilized and how proteins can be additionally stabilized are described with special emphasis on the use of amino acid substitutions by site-directed mutagenesis.

The folding of a polypeptide chain is driven by the strong tendency of sequestering hydrophobic side chains from solvent, which are followed or accompanied by various non-covalent interactions; i.e. H-bonding between peptide bonds (secondary structure) and the interactions, such as H-bonding, electrostatic interactions, hydorphobic interactions, and van der Waals contacts, between side chains and between main chain and side chains (see Chapter 2). The three-dimensional structure is finally formed with a delicate balance between the stabilizing factors derived from these non-covalent interactions and the destabilizing factors due to the conformational entropy between the folded and unfolded forms of the protein. It is considered that protein structure or protein stability is principally maintained by some *key* interaction(s). Matsumura *et al.* (1988), for example, reported that the local hydrophobic effect around position 3 in T4 lysozyme may contribute to the overall stability of the enzyme. The disruption of the key interaction(s) by heat or denaturnants such as urea and guanidine hydrochloride will result in cooperative unfolding of the protein (Tanford, 1968). The results so far accumulated on the stability and stabilization of proteins appear to indicate that the strategy for stabilizing proteins, i.e. the key interactions, differ from protein to protein. Any generalized mechanisms for protein stability have not yet been presented (see Chapters 3 and 6).

To reveal the key interaction(s) or the participated amino acid residues, many studies have been made (see Chapters 3–5). As described in Chapters 4 and 5, thermophilic proteins which are more stable as compared to their mesophilic counterparts, and mutant proteins of different stability relative to its wild-type protein, were used as materials for that purpose (Nosoh and Sekiguchi, 1988). However, from comparison of protein structures, especially the amino acid sequence, between thermophilic and mesophilic proteins, any molecular reasons for the enhanced stability of thermophilic proteins, and therefore for protein stability, could not be deduced, although some valuable guidelines for engineering proteins were proposed (e.g. Menendez-Arias and Argos, 1989). Mutant proteins, on the other hand, can afford much more better samples for analyzing protein stability (Yutani et al., 1977; Shortle and Meeker, 1986; Grutter et al., 1987). Recent developments in recombinant DNA technique have made it possible to substitute any amino acid(s) at any position in proteins at will by site-directed mutagenesis (Ulmer, 1983). Through this procedure (protein engineering) one can estimate the correlation of amino acid substitution with stability change, and thus the relation of non-covalent interaction(s) to protein stability (see Chapters 3 and 6). Protein engineering is thus becoming a promising strategy for analyzing protein stability as well as for stabilizing proteins (Fig. 7.1). In addition, through protein engineering enzymes can be stabilized even for non-aqueous solvents (Arnold, 1990), as described later.

Estimation of stability change with amino acid substitution

It is indeed an easy task to substitute any amino acid(s) in proteins by site-directed mutagenesis, provided that the gene coding for the protein is available. For analyzing

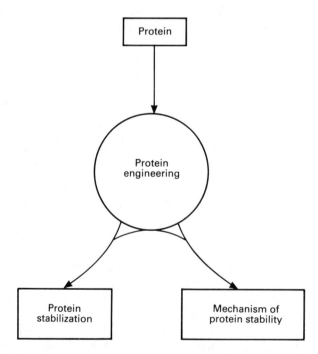

Fig. 7.1—Relationship of amino acid substituation to controlling protein stability and dissecting its mechanism. Proteins are stabilized through site-directed mutagenesis based on the mechanism of protein unfolding. Site-directed mutagenesis is also employed to reveal the mechanism of stabilization (e.g. comparison of X-ray crystal structures and thermodynamic and kinetic evulation in the wild-type and engineered proteins).

protein stability the contribution of the amino acid substitution (or the resulted interaction change) to the altered protein stability should be quantitatively evaluated. Nowadays, the difference ($\Delta\Delta G$) between the free energy change for protein unfolding (ΔG) of the mutant and wild-type proteins is most commonly used for the quantitative estimate of stability change. As described in section 6.2, the candidate sites for amino acid substitution are selected from model building using coordinates from a refined crystal structure of the protein to be engineered, ascertaining that neither conformational change around the site nor overall structure of the engineered protein occur. Due to the limitation of accuracy in model simulation, however, any unexpected conformational changes may occur after mutation. Actually, on replacing Gly with Ala in T4 lysozyme (Matthews et al., 1987), it was confirmed by X-ray crystal analysis that no change in whole structure but localized conformational adjustments in the vicinity of the substituted site had occurred. The unexpected changes in stability of subtilisin on Cys or disulfide bond introduction was considered to be due to the disruption of pre-existing interaction(s) (Wells and Powers, 1986; Mitchinson and Wells, 1989) and to some deviation from an idealized fit of an introduced disulfide bond (Pantoliano et al., 1987). On replacing Ile with smaller aliphatic amino acids in ribonuclease the decreased stability is reported to be due to the formation of a cavity in the protein interior (Kellis et al., 1988). The three-dimensional structures of the wild-type and mutant proteins should be compared in detail by X-ray crystallographic analysis. If no conformational change occurs on mutation, the change in protein stability can be directly correlated with the amino acid substitution. In the case of conformational change, the interpretation of the effect of the substitution on protein stability will become complex.

Structural determinations

For X-ray analysis large, perfect crystals (Wyckoff et al., 1985) are needed. Usually proteins are crystallized at atmospheric pressures by adding precipitant such as salts (e.g. ammonium sulfate) and alcohols (e.g. ethanol), which reduce their solubility. Crystallization is often a very slow process and is still considered to be difficult and time-consuming procedure (Delucas and Bugg, 1987). Recently, Visuri et al. (1990) succeeded in crystallizing glucose isomerase at high pressure, which stimulates supersaturation (Osugi et al., 1969). Supersaturation is needed for nucleation and growth of crystals. At 1500 bar crystal formation starts after several minutes and after 30 min uniform but small crystals are formed in amounts of 40–50 g l^{-1}. Although more work is necessary, pressure crystallization appears to open new and interesting perspectives for protein structural studies.

The nature of the compact globular states of proteins has been well studied by X-ray diffraction methods with protein crystals (Richardon, 1981), and recently with nuclear magnetic resonance (NMR) techniques it became possible to directly determine the structure of proteins in solution (Wuthrich, 1986, 1989). The method was principally based on three elements: (1) nuclear Overhauser enhancement experiments which enable measurements of ^1H–^1H distances in macromolecules (Wagner and Wuthrich, 1979); (2) an efficient technique for obtaining sequence-specific assignments of the many hundred to several thousand lines in a protein (Billeter et al., 1982; Wagner and Wuthrich, 1982; Wider et al., 1982); and (3) a distance geometry algorithm (Crippen and Havel, 1981) specifically implemented for computing three-dimensional protein structures from NMR data (Braun et al., 1981). Of these three elements, ^{1}H–^{1}H distance measurements had became possible before resonance assignment were available. The sequential resonance assignment technique (1) and three-dimensional protein structure determined by NMR and distance geometry (3) were developed and applied before two-dimensional NMR experiments were ready for reliable measurements with proteins. The situation is similar to that of protein crystallography at the time before the heavy-atoms derivative technique was presented for solving the phase problem (Wuthrich et al., 1982). However, the greatly improved resolution and efficiency afforded by two-dimentional NMR spectroscopy (Wuthrich, 1986; Ernst et al., 1987) made it possible to determine the complete structure of a proteinase inhibitor in solution (Wuthrich et al., 1984).

From both diffraction and spectroscopic techniques and from theoretical studies the dynamic properties of the folded states have also been reavealed (McCammon and Karplus, 1983). However, very little is known about the unfolded or partially folded states of proteins. Crystallization of proteins in such states seems difficult or impossible, but NMR spectroscopy seems to be suited to their characterization. NMR spectroscopy can provide both structural and dynamic information about macromolecules in solution, although the method requried for such studies may be different from those now becoming familiar from studies of proteins in globular states, and the nature of any conformational description may need to be significantly different from that used for the globular states, because of the greater conformational freedom likely to be characterstic of the unfolded or partially folded states.

Recently, the detailed comparison of the unfolded form with the folded form of proteins was performed with hen-egg-white lysozyme (Dobson et al., 1987), α-lactoalbumin (Baum et al., 1989), staphylococcal nuclease (Evans et al., 1989) and urokinase (Oswald et al., 1989), and conformational studies of transition intermediate on reversible unfolding of the latter three proteins were made (Baum et al., 1989; Evans et al.,1989; Oswald et al., 1989). The structural heterogeneity in staphylococcal nuclease, due to Pro 117, was confirmed by the amino acid substitution, Pro 117 to Gly (Evans et al., 1987). The structural analysis of protein denaturation by the NMR technique, although in an early stage, may afford some valuable information in understanding the mechanism of protein stability which the X-ray technique is unable to afford. Therefore, NMR analysis might provide a useful technique, together with X-ray crystal analysis, for protein engineering.

Estimation of protein stability

As described in Chapters 3 and 6, a promising strategy for analyzing protein stability is to reveal how individual interactions, and therefore the partcipating amino acid residues in the interactions, contribute to stabilizing proteins through site-directed mutagenesis. To evaluate quantitatively these contributions, the alteration in protein stability by amino acid substitutions should be determined by the difference ($\Delta\Delta G$) in free energy change (ΔG) for protein unfolding between the mutant and wild-type proteins.

For many monomeric globular proteins, as stated in Chapter 3, a two-state mechanism (Lumry et al., 1966) is assumed for their reversible denaturation; i.e. at equilibrium between the folded (native) and unfolded (denatured) forms the concentration of partially folded intermediate molecules is small enough to be neglected. The value of (ΔG) for the urea or guanidine hydrochloride-induced denaturation can be determined by estimating the equilibrium constant between the folded and unfolded forms; e.g. the denaturation of ribonuclease A (Pace et al., 1990). In contrast to detergent-induced protein unfolding, the analysis of thermal unfolding appears not so simple but rather complicated.

As described in section 3.1.2, the value of ΔC_p is necessary for calculating ΔG. The value of C_p of unfolded protein is greater than that of folded protein, which gives rise to the large positive value of ΔC_p. If ΔH can be measured as a function of temperature, ΔC_p can be calculated by the equation $d(\Delta H)dT = \Delta C_p$. This method has been successful in some cases (Pace and Tanford, 1968; Jackson and Brandts, 1970). Another method is to measure T_m (Chowdhry and Colem, 1989) and ΔH_m at T_m as a function of pH, and to obtain ΔG from the slope of a plot of ΔH_m versus T_m. If proper precautions are taken, and if ΔH_m and ΔC_p do not depend on pH, an estimate of ΔC_p is possible (Privalov, 1979; Becktel and Schellman, 1987). In this method a differential calorimeter is used to dertermine T_m. Becktel and Schellman (1987) proposed the temperature of maximum stability, T_s. For ribonuclease T1, T_s occurs near $-5°C$. The value of T_s generally falls between $-10°C$ for a relatively hydrophilic protein like ribonuclease A and $35°C$ for a relatively hydrophobic protein like β-lactoglobulin. Thus, the heat or cold-induced unfolding of a protein like β-lactoglobulin can be directly observed (Pace and Tanford, 1968). A study on cold denaturation of T4 lysozyme has been reported (Chen and Schellman, 1989). Recently Pace and Laurents (1989) proposed a new method for determining ΔC_p. At temperatures below T_m they measured ΔG using urea-induced denaturation curves at varying temperatures. The values of ΔG above T_m were calculated from the thermal denaturation curves.

The unusual temperature dependence of protein folding depends on ΔC_p which is the difference between C_p (unfolded state) and C_p (folded state), and ΔC_p depends on the ordering of water molecules around the non-polar groups that are brought into contact with water when proteins unfold (Baldwin, 1986; Privalov and Gill, 1988). In order to calculate ΔG at some reference temperature for comparing the relative stabilities of a set of mutant proteins, ΔC_p should be determined for each one. The value of ΔC_p of wild-type protein can be used in the calculation (Becktel and Schellman, 1987). However, this is not always a reliable assumption, because the ΔC_p values of the mutant proteins may be greatly different from that of wild-type protein, as observed for the mutant proteins of staphylococcal nuclease (Shortle et al., 1988).

Kellis et al. (1988) claimed that urea-induced denaturation is better for measuring the ΔG for reversible denaturation of barnase than thermal denaturation (see section 6.2.2.1). Thermal denaturation is not adequate for comparing quantitatively the values of ΔG of different proteins with different T_m (Creighton, 1983), and a more completely unfolded structure can be produced by urea denaturation than thermal denaturation (Creighton, 1983; Pace, 1986). In addition, as described above, the problem on ΔC_p should also be considered for thermal denaturation.

Thermodynamic and kinetic stability

Protein denaturation generally proceeds according to the following equation:

$$N \rightleftharpoons D \rightarrow I$$

where N, D and I represent the folded (native), reversibly unfolded, and irreversibly denatured forms of proteins, respectively (Klibanov, 1983). When the second process is extremely slow as compared to the first process, protein denaturation is reversible, and thermodynamic analysis can be performed for the reversible unfolding reaction, as exemplified with small, monomeric globular proteins (Chapters 3 and 6). For most globular proteins, especially oligomeric, however, reversible unfolding is followed by the second, irreversible process. Thus, for irreversible denaturation thermodynamics cannot be applied, and kinetic analysis is used. The possible mechanisms of protein denaturation are summarized by Mozheav *et al.* (1988).

As stated with many examples (Chapter 6), the stability of proteins exhibiting reversible denaturation is most commonly expressed by the free change on protein unfolding (see section 3.1.2). If denaturation is not a simple reversible transition between monomeric folded and unfolded states, as in λ repressor (Hecht *et al.*, 1984), however, the accurate determination of thermodynamic parameters becomes impossible (Hecht *et al.*, 1986). For most irreversible denaturation processes, on the other hand, the reaction constants for irreversible denaturation, e.g. thermal inactivation of hen egg-white lysozyme (Ahern and Klibanov, 1985) and of the disulfide-introduced subtilisin (Mitchinson and Wells, 1989), and the free energy change between the native and activation (transition) states (Eyring, 1935), e.g. thermal denaturation of tryptophan synthase α-subunit (Yutani *et al.*, 1987) and of modified lactate dehydrogenase (Shibuya *et al.*, 1982), are often employed.

The stabilizing effects exerted on the equilibrium between the folded and unfolded states, which are exempilfied in Chapter 6, are often called thermodynamic stabilization. Thermodynamic stabilization can be performed by increasing the stabilizing factors (section 6.2.2) and by decreasing the destabilizing factors (section 6.2.1). Kinetic stabilization is referred to as stabilization against irreversible inactivation (the conversion of unfolded state to irreversibly changed state). When the irreversible denaturation of proteins is triggered by, for example, the oxidation of SH (Brot and Weissbach, 1983; Ahern and Klibanov, 1985), thiol/disulfide interchange (Wang *et al.*, 1984), oxidation of Met (Stauffer and Eston, 1969) and deamidation of Asn (Ahern and Klibanov, 1985), replacing these amino acid residues by other stable ones may prevent the denaturation (Perry and Wetzel, 1986; Estell *et al.*, 1985; Ahern *et al.*, 1987). The modification of the unfolded state to more compact structures, as shown in the disulfide bonded T4 lysozyme (Wetzel *et al.*, 1988), appears impossible to design for protecting against irreversible denaturation.

Amino acid substitutions

Proteins are stabilized by the delicate balance between the stabilizing factors, such as hydrophobic interactions, electrostatic interactions, H-bonding and van der Waals contacts, and destabilizing factors of entropic effect on protein unfolding (Chapter 3). Therefore, for designing to stabilize proteins and for analyzing protein stability, enhancing the stabilizing factors or reducing the destabilizing factors has been attempted, as described in Chapters 3 and 6.

The hydrophobic interior of proteins consists mostly of tightly packed side chains, many of them aliphatic amino acids, and their arrangement defines the relationships between large units of secondary and tertiary structure (Chothia, 1984). Thus, interior packing determines much of the overall shape and the arrangements of the residues on the surface of proteins (Lesk and Chothia, 1980; Ponder and Richards, 1987). It may therefore be possible by rearranging protein interiors to alter active sites (Bone et al., 1989) and to stabilize proteins (Wetzel, 1987).

In understanding how to pack protein interiors there are two approaches based on different properties of the protein core. One is the liquid-like, hydrophobic properties of the core (Nozaki and Tanford, 1971; Tanford, 1980; Radzicka and Wolfenden, 1988). According to this model, replacing small aliphatic residues with larger ones in the protein interior stabilizes proteins by the increased hydrophobicity of the replaced side chains (see Chapters 3 and 6). Through improving the hydrophobicity of the protein interior some proteins have been stabilized (e.g. Yutani et al., 1987; Matsumura et al., 1988). However, it was recently shown that the hydrophobicity of proteins with interior aliphatic amino acid subtitutions does not always correlate to the hydrophobicity of the substituting residues (Sangberg and Terwilliger, 1991). The stability of some proteins increases proportionally with increasing hydrophobicity of substituted aliphatic residues, but with different slopes. The proteins and substitution sites employed for this analysis are tryptophan synthase α-subunit with substitution at position 49 (Yutani et al., 1987), barnase substituted at positions 88 and 96 (Kellis et al., 1989) and bacteriophage f1 gene protein (gVp) substituted at position 47 (Sandberg and Terwilliger, 1991). Other proteins, however, are less stabilized or destabilized with increased hydrophobicity of aliphatic residues; T4 lysozyme replaced at positions 3 and 129 (Matsumura et al., 1988; Karpusas et al., 1989), diydrofolate reductase substituted at position 75 (Garvey and Metthews, 1989) and gVp substituted at position 35 (Sandberg and Terwilliger, 1991). In addition, it was reported that the arrangement of buried residues, not their composition, is important in determining protein stability (Sandberg and Terwilliger, 1989). These results suggest that there exists no simple correlation between the hydrophobicity of interior residues and stability of proteins.

Another property of the protein interior is their densely packed, solid state (Richards, 1977). Close packing of interior atoms means that a side chain can make favorable van der Waals contacts with more atoms than it might in a liquid hydrocarbon. This produces the packing energy (Richards, 1977). Thus, on analyzing the effect of interior aliphatic amino acid substitutions on protein stability, Sandberg and Terwilliger (1991) emphasized the contributions of packing energy to protein stability, as well as the hydrophobic stabilization. The packing energy reflects an energetic favor or disfavor for introducing a large aliphatic side chain at an interior site previouly occupied by Ala. If there is not enough room at the site for larger residues, replacing Ala with larger side chains is energetically unfavorable, and it will exert a destabilizing effect. The smaller than expected stabilization in mutant proteins of dihydrofolate reductase Ala at position 75 (Garvey and Matthews, 1989) and of gVp at position 35 (Sandberg and Terwilliger, 1991) can be explained by such a packing energy effect.

The packing energy is the free energy difference between placing a large aliphatic side chain in the interior of a protein and placing it in a liquid hydrocarbon (Sandberg and Terwilliger, 1989). Sandberg and Terwilliger (1991) define a 'unit volume packing energy' (A) for a given site as the ratio of the packing energy (the difference in packing energy between a large substituent and Ala) to the difference in their volume. Sandberg and Terwilliger (1991) proposed a strategy for stabilizing proteins. The first step is to identify buried apolar residues, because increases in hydrophobicity at these sites may produce the largest stabilization. Any two mutant proteins with aliphatic side chains of different sizes are constructed to calculate a rough estimate of the value of A. One or two substitutions by site-directed mutagenesis will suffice for each site to be analyzed. The effects on stability of further substitutions can be predicted by the combination of the value of A with the anticipated hydrophobicity increase. In this way all the possible interior sites in a protein can be assessed for stabilization without exhaustive mutagenesis of the protein. This strategy is useful for indentifying sites where large aliphatic side chains will strongly destabilize or stabilize a protein, or sites where the stability is independent of the nature of the amino acid side chain and any apolar residue is tolerated.

Engineering proteins for non-aqueous media

The experimental and theoretical analyses of the stability and stabilization of proteins described in this book are all performed with the proteins in aqueous solvents. As mentioned in Chapters 2 and 3, the native conformation of proteins is maintained by a delicate balance of various non-covalent interactions, and water participates in all of them. Removal of water interacting with proteins by water-missible (polar) organic solvents will result in conformational change of proteins, due to forming new interactions between polar residues on the protein surface; e.g. protein denaturation caused by organic solvents such as enthanol and acetone. In non-polar organic solvents containing only small amount of water, however, proteins may still retain native conformations and exhibit activity (Klibanov, 1986; Laane *et al.*, 1987; Halling, 1987; Dordick, 1989), especially if some rules are followed (Klibanov; 1986, 1989). The rules are: (1) enzymes to be used in organic solvents should be lyophilized from aqueous solutions of pH optimal for enzyme activity; thereby the ionic groups of the protein can maintain the ionization states corresponding to the pH of the aqueous solution, even in organic solvents (pH memory); and (2) organic solvents should be strongly hydrophobic, as demonstrated by Zaks and Klibanov (1988a): the activity of chymotrypsin decreased with increasing hydrophilicity of organic solvents. Monomeric enzymes such as chymotrypsin and subtilisin in anhydrous non-polar organic solvents (e.g. octane) contain, on average, less than 50 molecules of bound water (Zaks and Klibanov, 1988a). To form a monolayer on the surface of such enzymes some 500 molecules of water are required (Rupley *et al.*, 1983). The water molecules bound to proteins probably form a few clusters around charged groups on the proteins. Oligomeric enzymes such as alcohol dehydrogenase, tyrosinase and alcohol oxidase are catalytically active in organic solvents when several hundreds of water molecules are bound per protein molecules (Zaks and Klibanov, 1988a). The levels of bound water molecules may form a monolayer coverage.

Thermal denaturation of proteins requires ample conformational mobility, and all reactions causing irreversible thermal inactivation of enzymes involve free water (Ahern and Klibanov, 1985; Zale and Klibanov, 1986). Thus both processes can be prevented by dehydration. Indeed, the half-life of chymotrypsin at 100°C was several hours in anhydrous octane, whereas in water even at 60°C the enzyme irreversibly inactivated within minutes (Zaks and Klibanov, 1988a). Thermostability depends on the nature of the solvent and the pH of the aqueous solution from which the enzyme is lyophilized. Subtilisin (Russell and Klibanov, 1988) and lipase (Zaks and Klibanov, 1984) were also stable for hours at 100°C. Lipase is not only stable but also active even at 100°C. This indicates that, in addition to enhanced thermal resistance, the enzyme still holds its folded state under such extreme conditions. Enzymes in organic solvents can be stored for a much longer period as compared to aqueous solution. Chymotrypsin, for example, is completely stable for six months in anhydrous octane at 20°C, while in water its half-life is only a few days (Zaks and Klibanov, 1988a).

Appilcations of enzymes in water are limited by the poor solubility in that medium of most organic substrates. Thus, enzyme reactions which are impossible in aqueous media may become possible in non-aqueous solvents. Highly non-polar solvents, however, are often poor solvents for substrates of interest. In addition, enzymes are in a dry or partially hydrated form in non-polar organic solvents (Zaks and Klibanov, 1988a). Enzymes in a soluble state, as in water or polar organic solvents, must be more active than in anhydrous non-polar organic solvents. However, there is a correlation between the loss of enzyme activity in non-polar solvents and solvent polarity (Lanne *et al.*, 1985; Reslow *et al.*, 1987; Zaks and Klibanov, 1988a). Thus, engineering enzymes for polar organic solvents has been attempted (Arnold, 1990).

Enzymes which are active and stable in organic solvents can be designed on the molecular basis of protein stability in aqueous solution:

(1) Protein surface–environment compatibility: even at extremely low levels of hydration (Zaks and Klibanov, 1988a), water solvates charged residues on the protein surface and mediates proton redistribution to establish the normal order of the pK_a values of these residues (Careri *et al.*, 1980). One approach for stabilizing proteins in respect to surface-charged residues is to reduce the amount of water by fixing the original pK_a values in water and removing water from non-polar organic solvents (Klibanov, 1989). Another is to reduce the hydration of charged sites by selectively eliminating charged residues from the protein surface through site-specific mutagenesis.

(2) Conformational stability: basic conformation of the interior of proteins stable in non-aqueous solvents may not differ from that of proteins stable in water. The strategy for improving conformational stability of proteins in aqueous solvents can be expected to apply to proteins in non-aqueous solvents (see section 6.2.1.2). Crambin is soluble and stable in polar organic solvents (DeMarco *et al.*, 1981; Arnold, 1989). Although crambin is insoluble in water, the protein has several water-soluble homologs with similar folded structures (Teeter *et al.*, 1981). Crambin will be a good model for protein engineering for stability in organic solvents.

(3) Disulfide bonds: disulfide cross-links stabilize proteins in aqueous solvent by decreasing the entropy of the unfolded form of proteins (see section 6.2.1.1). The introduction of disulfide bonds may also stabilize proteins in non-aqueous solvents, as in the case of conformational stability.

(4) Hydrophobic interactions: proteins in aqueous solution are folded and stabilized

by hydrophobic interactions, i.e. by the change in transfer free energy from water to ethanol (a model of the protein interior) (see section 2.3.1.5, 3.2.3 and 6.2.2.1). The hydrophobic contribution to protein stability will be greatly reduced in non-aqueous solvents, because the magnitudes of transfer free energy going from non-aqueous solvent to ethanol are small. However, strengthening tight packing of the protein interior and van der Waals contacts between non-polar side chains may be expected to stabilize proteins in non-aqueous solvents through hydrophobic amino acid substitutions. Transfer free energies do not correctly evaluate the effects of packing of the protein interior and van der Waals contacts, and thus cannot account for all the effects of amino acid substitution on proteins stability (Privalov and Gill, 1988).

(5) H-bonds: surface-exposed H-bonding sites are solvated in aqueous solution, whereas they can form no or only weak bonds with non-aqueous solvents. This may lead to the formation of alternative, inactive conformation of proteins that result in more H-bonding internally. When one uses non-aqueous solvents for proteins, the residues with H-bonding potential are removed or satisfied through the addition of external H-bonding agents (Zaks and Klibanov, 1988a). The replacement of Asp at position 218 to Ser in subtilisin BPN' improved bond parameters for several existing H-bonds and stabilized the protein by about $1\,\mathrm{kcal\,mol^{-1}}$ at 58.5°C in aqueous solution (Bryan et al., 1986). Other subsitutions showed the stabilizing effects to be consistent with H-bonding properties. Introduction of such additional or more stable H-bond might also increase the stability of proteins in non-aqueous solvents.

(6) Electrostatic interactions: these are not stable in hydrophobic environments, e.g. in non-aqueous solvents, and therefore almost always are found in polar environments in folded proteins (Warshel et al., 1984). A solvent-accessible ion pair is shown to contribute $3-5\,\mathrm{kcal\,mol^{-1}}$ to the stability of T4 lysozyme (Anderson et al., 1990). Ion pairs in the protein interior, however, are stabilized by a relatively fixed complementary electrostatic environment of protein charges or H-bonds (Hwang and Warshel, 1988). The addition of ion pairs stabilized by protein dipoles may be one approach to stabilizing proteins in non-aqueous solvents.

Despite the widespread interest in enzymes in non-aqueous solvents, there are still only a few published works on the effects of amino acid substitutions on protein stability in non-aqueous solvents.

Pantoliano et al. (1989) produced a highly thermostable subtilisin BPN' mutant by combining six amino acid subsititutions. The mutant protein is indeed 100 times more stable than the wild-type protein, with no change in enzyme activity. The amino acid replacements and the presumed stabilization mechanisms in aqueous solution are: (1) Asn 218 to Ser, which improves H-bonding; (2) Gly 169 to Ala, which confers hydrophobic stabilization or alters the conformational entropy of the unfolded form; (3) Met 50 to Phe, which confers hydrophobic stabilization; (4) Tyr 217 to Lys, which forms H-bonding ; (5) Glu 206 to oxidized Cys, which improves van der Waals contacts or hydrophobic stabilization; and (6) Asn 76 to Asp, which improves Ca^{2+} Binding and H-bonding.

The mutant subtilisin was also more stable in anhydrous dimethylformamide (DMF) than the wild-type protein. In 50% DMF, however, no remarkable difference in stability was observed between the two proteins (Wong et al., 1990). The effects of individual amino acid substitutions have not reported on the stability of the subtilisin in DMF. The study of each individual mutation, however, is important for understanding their contributions to the stability of proteins in non-aqueous solvents.

To examine the design rules for stabilizing proteins in non-aqueous solvents as described above, Arnold and co-workers investigated the effects of amino acid substitutions on subtilisin E, which is homologous to subtilisin BPN' (Arnold, 1990).

They selected two substitutions: Asp 248 to Asn and Asn 218 to Ser. The substitution Asp 248 to Asn replaces a charged residue on the protein surface with an uncharged residue, thus rendering the surface less hydrophilic and less dependent on hydration. The other mutation, Asn 218 to Ser, is known to improve H-bonding in subtilisin BPN' and to significantly increase thermostability in aqueous solution (Pantoliano *et al.*, 1989). The same substitution was then introduced into subtilisin E to test the hypothesis that improved H-bonding will stabilize proteins in non-aqueous solvents. A double mutant containing both mutations was also examined.

The removal of single surface charge, Asp 248 to Asn, exhibited essentially no effect on the stability of subtilisin E in aqueous solution (Fig. 7.2a). The substitution of Asn 218 to Ser, which improves H-bonding, however, considerably stabilized the enzyme, and the substitutions showed an additive effect. In 40% DMF, as expected from the design rules, the Asp 248 to Asn mutation resulted in a small increase in stability, as compared to the wild-type protein (Fig. 7.2b). This appears to indicate the effect of surface charge on protein stability in non-aqueous solvents. The replacement of Asn 218 with Ser was more effective for stabilizing subtilisin E in 40% DMF than the Asp 248 to Asn substitution. This may support the effect of improving H-bonding on protein stability in non-aqueous solvents. As shown in the figure, the effects of the substitutions are approximately additive. Although much more evidence is needed, the results may support the idea that surface properties can influence protein stability in non-aqueous solvents.

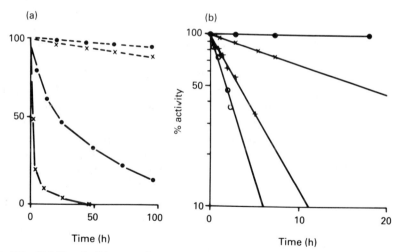

Fig. 7.2—Stability of wild-type and mutant subtilisins in non-aqueous solvent. (a) Stability of wild-type (×) and engineered (●) subtilisin BPN' in 50% DMF solution (---) and in ahydrous DMF (——). The engineered subtilisin contains six amino acid substitutions (Pantoliano *et al.*, 1988.) (b) Stability of wild-type and engineered subtilisin E in 40% DMF solution. (○) wild-type; (+) Asp 248 to Asn; (×) Asn 218 to Ser, (○) Asp 248 to Asn plus Asn 218 to Ser (cited from Arnold, 1990.)

Future directions

The stability and stabilization of proteins through protein engineering are progressing rapidly. From the results so far reported, however, it is apparent that protein engineering is a promising strategy for analyzing protein stability as well as for stabilizing proteins. Protein engineering for stabilizing proteins is based on the knowledge of the mechanism of protein stability, and the results on the stability change by amino acid replacements should grealty enhance the information on protein stability. Indeed, these two approaches go together.

In analyzing protein stability through protein engineering, the amino acid substitutions should be correctly or quantitatively estimated by the parameters representing protein stability. To date the most reliable parameters for protein stability are the thermodynamic parameters for reversible unfolding of proteins, i.e. ΔG (and its components, ΔH and ΔS). Since proteins are marginally stabilized, values of ΔG should be accurately determined to compare them between the mutant and wild-type proteins. As seen with λ repressor (Hecht et al., 1984), even a small monomeric protein does not always exhibit an ideal two-state unfolding. The conditions for ideal reversible unfolding for proteins should be set up. In estimating the value of ΔG attention should be paid to the role of water in protein unfolding (Shortle et al., 1988), and therefore ΔC_P should not be neglected for determining ΔG for thermal reversible denaturation (Pace, 1990). Disrupted or newly formed interactions or contacts in engineered proteins can be determined by precise X-ray crystallographic measurements. In future NMR measurements will also become useful procedure for detecting such conformational changes, and in addition this technique migth be useful for analyzing the protein unfolding process, as described in this chapter. Relating to structural determination, procedures for predicting secondary and also tertiary structures will be further developed, and will become used effectively when it is attempted to engineer proteins whose tertiary structure is not known.

Amino acid substitutions at all possible sites in a protein, together with detailed analyses of the structural and energetic changes on mutation, will produce many important data on protein stability, from which any conclusion on the mechanism of stability of the protein can be drawn. Thus, intensive studies with a small monomeric protein such as T4 lysozyme have been in progress (Chapters 3 and 6), because it is relatively easy to determine its three-dimensional structure, as compared to proteins of larger molecular sizes, and in general it undergoes reversible unfolding which makes it possible to quantitatively estimate the change in stability. Other small proteins such as subtilisin, nuclease and λ repressor are also sutiable targets for protein engineering. Combining together the results on different proteins it may be possible to determine whether proteins are stabilized by the same or other single interactions, or by the accumulation of many small changes throughout the whole protein molecule (Menendez-Arias and Argos, 1989).

Proteins denaturation at elevated temperature is often accompanied by an irreversible change in protein structure (Klibanov, 1983). If the irreversible denaturation is triggered by the irreversible thermal destruction of some amino acid residue in a protein, e.g. deamidation of Asn in yeast triose-phosphate isomerase (Ahern et al., 1987), the protein can be stabilized by replacing the heat-labile amino acid residue with other stable ones. Elucidation of the mechanism of thermal irreversible denaturation should be attempted with more proteins to stabilize them through protein engineering. Thermal inactivation of yeast triose-phosphate isomerase is caused by the deamidation of Asn located between subunit–subunit interfaces (Ahern *et al.*, 1987). On designing to stabilize oligomeric proteins, then, the subunit interaction should also be considered as well as the structure of each subunit molecule. It seems worthwhile attempting to stabilize an oligomeric protein, whose three-dimensional structure is known, by improving the stability of each subunit protein and also the subunit interactions. These approaches will greatly help in elucidating the mechanism of stability of oligomeric proteins.

To date, the largest increase in stability through a single amino acid substitution was reported with yeast iso-1-cytochrome c, which exhibits a 4.2 kcal mol^{-1} increase in stability as compared to the wild-type protein (Das *et al.*, 1989). The degree of stabilization of other proteins by site-directed mutagensis was much less than that with the yeast protein (see Chapter 6). For example, each of six substitutions in subtilisin increases the stability only by 0.3–1.3 kcal mol^{-1} (Pantoliano *et al.*, 1989). When the substitutions are combined, the stability increases by 3.8 kcal mol^{-1}. The additive effect of substitution was reported by Matsumura *et al.* (1988). Multi-substitutions at appropriate sites in a protein will increase its stability up to a desired level.

In the near future the stabilization of proteins in extreme environments, such as low temperatures, low and high pH values, high salt concentrations and high pressure, will be attempted through site-directed mutagenesis. The results obtained with these approaches will provide more interesting information on protein stability.

REFERENCES

Ahern, T. J. and Klibanov, A. M. (1985) The mechanism of irreversible enzyme inactivation at 100°C. *Science*, **228**, 1280–1284

Ahern, T. J., Casal, J. I., Petsko, G. A. and Klibanov, A. M. (1987) Control of oligomeric enzyme thermostability by protein engineering. *Proc. Natl. Acad. Sci. USA*, **84**, 675–679.

Anderson, D. E. Becktel, W. J. and Dahlquist, F. W. (1990) pH-induced denatutaion of proteins: a single salt bridge contributes 3–5 kcal/mol to the free energy of folding of T4 lysozyme. *Biochemistry*, **29**, 2403–2408.

Arnold, F. H. (1989) NMR studies of crambin structure and unfolding in nonaqueous solvents. *Ann. NY Acad. Sci.*, **542**, 30–36.

Arnold, F. H. (1990) Engineering enzymes for non-aqueous solvents. *Trends Biotechnol.*, **8**, 244–249.

Baldwin, R. L. (1986) Temperature dependence of the hydrophobic interaction in protein folding. *Proc. Natl. Acad. Sci. USA*, **83**, 8069–8072.

Baum, J., Dobson, C. M., Evans, P. A. and Hanley, C. (1989) Characterization of a partly folded protein by NMR methods: studies on the molten globule state of guinea-pig α-lactoalbumin. *Biochemistry*, **18**, 7–13.

Becktel, W. J. and Schellman, J. A. (1987) Protein stability curves. *Biopolymers*, **26**, 1859–1877.

Billeter, M., Braun, V. J. and Wuthrich, K. (1982) Sequential resonance assignments in protein ^{1}H nuclear magnetic resonance spectra, computation of sterically allowed proton–proton distances and statistical analysis of proton–proton distances in single crystal protein conformations. *J. Mol. Biol.*, **155**, 321–346.

Bone, R., Silen, J. L. and Agard, D. A. (1989) Structural plasticity broadens the specificity of an engineered protease. *Nature (London)*, **339**, 191–195.

Braun, W., Bosch, C., Brown, L. R., Go, N. and Wuthrich, K. (1981) Combined use of porton–proton Overhauser enhancements and a distance geometry algorithm for determination of polypeptide conformations: application to micelle-bound glucagon. *Biochim. Biophys. Acta*, **667**, 377–396.

Brot, N. and Weissbach. H. (1983) Biochemistry and physiological role of methionine sulfoxide residues in proteins. *Arch. Biochem., Biophys.*, **223**, 271–281.

Bryan, P. N., Rollence, M. L., Pantoliano, M. W., Wood, J., Finzel, B. C., Gilliland, G. L. Howard, A. J. and Poulos, T. L. (1986) Proteases of enhanced stability: characterization of a thermostable variant of subtilisin. *Proteins Struct. Funct. Genet.*, **1**, 326–334.

Careri, C., Gratton, E., Yang, P.-H. and Ruppley, J. A. (1980) Correlation of IR spectroscopic, heat capacity, diamagnetic susceptibility and enzymic measurements on lysozyme powder. *Nature (London)*, **284**, 572–573.

Chen, B. L. and Schellman, J. A. (1989) Low-temperature unfolding of a mutant of phage T4 lysozyme. 1. Equilibrium studies. *Biochemistry*, **28**, 685–691.

Chothia, C. (1984) Principles that determine the structure of proteins. *Annu. Rev. Biochem.*, **53**, 537–572.

Chowdhry, B. Z. and Colem, S. C. (1989) Differential scanning calorimetry: applications in biotechnology. *Trends Biotechnol.*, **7**, 11–18.

Creighton, T. E. (1983) *Proteins*. Freeman, New York.

Crippen, G. M. and Havel, T. F. (1981) *Distance geometry and molecular conformation* Wiley, New York.

Das, G., Hickey, D. R., McLendon, D. McLendon, G. and Sherman, F. (1989) Dramatic thermostabilization of yeast iso-1-cytochrome C by an asparagine → isoleucine replacement at position 57. *Proc. Natl. Acad. Sci. USA*, **86**, 496–499.

Delucas, L. J. and Bugg, C. E. (1987) New directions in protein crystal growth. *Trends Biotechnol.*, **5**, 188–193.

DeMarco, A., Lecomte, J. T. J. and Llinas, M. (1981) Solvent and temperature effects on crambin, a hydrophobic protein, as investigated by proton magnetic resonance. *Eur. J. Biochem.*, **119**, 483–490.

Dobson, C. M., Evans, P. A., Fox, R. O., Redfield, C. and Topping, K. D. (1987) Two dimensional exchange experiments in NMR studies of protein dynamics and folding. *Protides Biol. Fluids*, **35**, 433–436.

Dordick, J. S. (1989) Enzymic catalysis in monophasic organic solvents. *Enzyme Microb. Technol.*, **11**, 194–211.

Ernst, R. R., Bodenhausen, G. and Wokaun, A. (1987) *Principles of nuclear magnetic resonance in one and two dimensions*. Clarendon, London.

Estell, D. A., Graycar, T. P. and Wells, J. A. (1985) Engineering an enzyme by site-directed mutagenesis to be resistant to chemical oxidation. *J. Biol. Chem.*, **260**, 6518–6521.

Evans, P. A., Kautz, R. A., Fox, R. O. and Dobson, C. M. (1987) Protein isomerism in staphylococcal nuclease characterized by NMR and site-directed mutagenesis. *Nature (London)*, **329**, 266–268.

Evans, P. A., Kautz, R. A., Fox, R. O. and Dobson, C. M. (1989) A magnetisation transfer NMR study of the folding of staphylococcal nuclease. *Biochemistry*, **28**, 362–370.

Eyring, H. (1935) Activated complex in chemical reactions. *J. Chem. Phys.*, **3**, 107–115.

Garvey, E. P. and Matthews, C. R. (1989) Effects of multiple replacements at a single position on the folding and stability of dihydrofolate reductase from *Escherichia coli*. *Biochemistry*, **28**, 2083–2093.

Grutter, M. G., Gray, T. M., Weaver, L. H., Alber, T., Wilson, K. and Matthews, B. M. (1987) Structural studies of mutants of the lysozyme of bacteriophage T4: the temperature-sensitive mutant protein Thr 157 to Ile. *J. Mol. Biol.*, **197**, 315–329.

Halling, P. J. (1987) Biocatalysis in multi-phase reaction mixtures containing organic liquids. *Biotechnol. Adv.*, **5**, 47–84.

Hecht, M. H., Sturtevent, J. M. and Sauer, R. T. (1984) Effect of single amino acid replacements on the thermal stability of the NH_2-terminal domain of phagse λ repressor. *Proc. Natl. Acad. Sci. USA*, **81**, 5685–5689.

Hecht, M. H., Sturtevent, J. M. and Sauer, R. T. (1986) Stabilization of γ repressor against thermal denaturation by site-directed Gly $\rightarrow$ Ala changes in α-helix 3. *Proteins Struct. Funct. Genet.*, **1**, 43–46.

Hwang, J.-K. and Warshel, A. (1988) Why ion pair reversal by protein engineering is unlikely to succeed. *Nature (London)*, **334**, 270–272.

Jackson, W. M. and Brandts, J. F. (1970) Thermodynamics of protein denaturation: calorimetric study of the reversible denaturation of chymotrypsinogen and conclusions regarding the accuracy of the two-state approximation. *Biochemistry*, **9**, 2294–2301.

Karpusas, M., Baase, W. A., Matsumura, M. and Matthews, B. W. (1989) Hydrophobic packing in T4 lysozyme probed by cavity-filling mutants. *Proc. Natl. Acad. Sci. USA*, **86**, 8237–8241.

Kellis, J. T., Nyberg, K., Jr, Sali, D. and Fersht, A. R. (1988) Contribution of hydrophobic interactions to protein stability. *Nature (London)*, **333**, 784–786.

Kellis, J. T., Nyberg, K. and Fersht, A. R. (1989) Energetics of complementary side chain packing in a protein hydrophobic core. *Biochemistry*, **28**, 4914–4922.

Klibanov, A. M. (1983) Stabilization of enzymes against thermal inactivation. *Adv. Appl. Microbiol.*, **29**, 1–28.

Klibanov, A. M. (1986) Enzymes that work in organic solvents. *Chemtech.*, **16**, 354–359.

Klibanov, A. M. (1989) Enzymatic catalysis in anhydrous organic solvents. *Trends Biochem. Sci.*, **14**, 141–144.

Laane, C., Boeren, S. and Vos, K. (1985) On optimizing organic solvents in multi-liquid-phase biocatalysis. *Trends Biotechnol.*, **3**, 251–252.

Laane, C., Boeren, S., Vos, K. and Veeger, C. (1987) Rules for optimization of biocatalysis in organic solvents. *Biotechnol. Bioeng.*, **30**, 81–87.

Lesk, A. M. and Chothia, C. (1980) How different amino acid sequences determine similar protein structures: the structure and evolutionary dynamics of the globins. *J. Mol. Biol.*, **136**, 225–270.

Lumry, R., Biltonen, R. and Brandts, J. F. (1966) Validity of the 'two-state' hypothesis for conformational transition of proteins. *Biopolymers*, **4**, 917–944.

Matsumura, M., Yasumura, S. and Aiba, S. (1986) Cumulative effect of intragenic amino acid replacements on the thermostability of a protein. *Nature (London)*, **323**, 356–358.

Matsumura, M., Becktel, W. and Matthews, B. W. (1988) Hydrophobic stabilization in T4 lysozyme determined directly by multiple substitutions of Ile 3. *Nature (London)*, **334**, 406–410.

Matthews, B. W., Nicholson, H. and Becktel, W. J. (1987) Enhanced protein thermostability from site-directed mutations that decrease the entropy of unfolding. *Proc. Natl. Acad. Sci. USA.*, **84**, 6663–6667.

McCammon, J. A. and Karplus, M. (1983) The dynamic picture of protein structure. *Accts. Chem. Res.*, **16**, 187–193.

Menendez-Arias, L. and Argos, P. (1989) Engineering protein thermal stability: sequence statistics point to residue substitutions in α-helices. *J. Mol. Biol.*, **206**, 397–406.

Mitchinson, C. and Wells, J. A. (1989) Protein engineering of disulfide bonds in subtilisin BPN'. *Biochemistry*, **28**, 4807–4815.

Mozhaev, V. V., Berezin, H. V. and Martinek, K. (1988) Structure–stability relationship in proteins: fundamental tasks and strategy for the development of stabilized enzyme catalysts for biotechnology. *CRC Crit. Rev. Biochem.*, **23**, 235–281.

Nosoh, Y. and Sekiguchi, T. (1988) Protein thermostability: mechanism and control through protein engineering. *Biocatalysis*, **1**, 257–273.

Nozaki, Y. and Tanford, C. (1971) Solubility of amino acids and two glycine peptides in aqueous ethanol and dioxane solutions: establishment of a hydrophobicity scale. *J. Biol. Chem.*, **246**, 2211–2217.

Osugi, J., Shimizu, K., Yasunami, K., Moritoki, M. and Onodera, A. (1969) Liquid–solid transition at high pressure III. Benzene, monochlorobenzene and toluene at 25°C. *Rev. Phys. Chem. Japan*, **38**, 90–95.

Oswald, R. E., Bogusky, M. J., Bamberger, M., Smith, R. A. G. and Dobson, C. M. (1989) Dynamics of multidomain fibrinolytic protein: a 2D NMR study of urokinase. *Nature (London)*, **337**, 579–582.

Pace, C. N. (1986) Determination and analysis of urea and guanidine hydrochloride denaturation curve. *Meth. Enzymol.*, **131**, 266–279.

Pace, C. N. (1990) Measuring and increasing protein stability. *Trends Biotechnol.*, **8**, 93–98.

Pace, C. N. and Tanford, C. (1968) Thermodynamics of the unfolding of β-lactoglobulin A in aqueous urea solution between 5 and 55°. *Biochemistry*, **7**, 198–208.

Pace, C. N. and Laurents, D. V. (1989) A New method for determining the heat capacity change for protein folding. *Biochemistry*, **28**, 2520–2525.

Pace, C. N., Laurents, D. V. and Thomson, J. A. (1990) pH dependence of the urea and guanidine hydrochloride denaturation of ribonuclease A and ribonuclease T1. *Biochemistry*, **29**, 2564–2572.

Pantoliano, M. W., Ladner, R. C., Bryan, P. N., Rollence, M. L., Wood, L. R. and Poulos, T. L. (1987) Protein engineering of subtilisin BPN': enchanced stabilization through the introduction of two cysteines to form a disulfide bond. *Biochemistry*, **26**, 2077–2082.

Pantoliano, M. W., Whitlow, M., Wood, J. F., Dodd, S. W., Hardman, K. D., Rollence, M. L. and Bryan, P. N. (1989) Large increase in general stability for subtilisin BPN' through incremental changes in the free energy of unfolding. *Biochemistry*, **28**, 7205–7213.

Perry, L. J. and Wetzel, R. (1986) Unpaired cysteine-54 interferes with the ability of an engineered disulfide to stabilize T4 lysozyme. *Biochemistry*, **25**, 733–739.

Ponder, J. W. and Richards, F. M. (1987) Tertiary templates for proteins: use of packing criteria in the enumeration of allowed sequences for different structural classes. *J. Mol. Biol.*, **193**, 775–791.

Privalov, P. L. (1979) Stability of proteins. *Adv. Protein Chem.*, **33**, 167–241.

Privalov, P. L. and Gill, S. J. (1988) Stability of Protein structure and hydrophobic intertaction. *Adv. Protein Chem.*, **39**, 191–234.

Radzicka, A. and Wolfenden, R. (1988) Comparing the polarities of the amino acids: side-chain distribution coefficients between the vapor phase, cyclohexane, 1-octanol, and neutral aqueous solution. *Biochemistry*, **27**, 1664–1670.

Reslow, M., Aldercreutz, P. and Mattiason, B. (1987) Organic solvents for bioorganic synthesis. I. Optimization of parameters for a chymotrypsin catalyzed process. *Appl. Microbiol. Biotechnol.*, **26**, 1–8.

Richards, F. M. (1977) Areas, volumes, packing, and protein stability. *Annu. Rev. Biophys. Bioeng.*, **6**, 151–176.

Richardson, J. S. (1981) The anatomy and taxonomy of protein structure. *Adv. Protein Chem.*, **23**, 167–339.

Rupley, J. A., Gratton, E. and Careri, G. (1983) Water and Globular proteins. *Trends Biochem. Sci.*, **8**, 18–22.

Russell, A. J. and Klibanov, A. M. (1988) Inhibitor-induced enzyme activation in organic solvents. *J. Biol. Chem.*, **263**, 11624–11626.

Sandberg, W. S. and Terwilliger, T. C. (1989) Influence of interior packing and hydrophobicity on the stability of a protein. *Science*, **245**, 54–57.

Sandberg, W. S. and Terwilliger, T. C. (1991) Repacking protein interiors. *Trends Biotechnol.*, **9**, 59–63.

Shibuya, H., Abe, M., Sekiguchi, T. and Nosoh, Y. (1982) Effect of guanidination on subunit interactions in hybrid isozymes from pig lactate dehydrogenase. *Biochim. Biophys. Acta*, **708**, 300–304.

Shortle, D. and Meeker, A. K. (1986) Mutant forms of stophylococcal nuclease with altered patterns of guanidine hydrochloride and urea denaturation. *Proteins Struct. Funct. Genet.*, **1**, 81–89.

Shortle, D., Meeker, A. K. and Freire, E. (1988) Stability mutants of staphylococcal nuclease: large compensating enthalpy–entropy changes for the reversible denaturation reaction. *Biochemistry*, **27**, 4761–4768.

Stauffer, C. E. and Eston, D. (1969) The effect on subtilisin activity of oxidizing a methionine residue. *J. Biol. Chem.*, **244**, 5333–5338.

Tanford, C. (1968) Protein denaturation. *Adv. Protein Chem.*, **23**, 121–282.

Tanford, C. (1980) *The hydrophobic effect*. Wiley, New York.

Teeter, M. M., Mazer, J. A. and L' Italien, J. J. (1981) Primary structure of the hydrophobic plant protein crambin. *Biochemistry*, **20**, 5437–5443.

Ulmer, K. M. (1983) Protein engineering. *Science*, **219**, 666–671.

Visuri, K., Kaipainen, E., Kivimaki, J., Niemi, H., Leisola, M. and Palosaari, S. (1990) A new method for protein crystallization using high pressure. *Biotechnology*, **8**, 547–549.

Wagner, G. and Wuthrich, K. (1979) Truncated driven nuclear Overhauser effect (TOE), a new technique for studies of selective ^1H–^1H Overhauser effects in the presence of spin diffusion. *J. Magn. Reson.*, **33**, 675–680.

Wagner, G. and Wuthrich, K. (1982) Sequential resonance assignments in protein ^1H nuclear magnetic resonance spectra: basic pancreatic trypsin inhibitor. *J. Mol. Biol.*, **155**, 347–366.

Wang, A., Lu, S. D and Mark. D. F. (1984) Site-directed mutagenesis of human interleukin-2-gene: structure–function analysis of the cysteine residues. *Science*, **224**, 1431–1433.

Warshel, A., Russell, S. T. and Churg, A. K. (1984) Macroscopic model for studies of electrostatic interactions in proteins: limitations and applicability. *Proc. Natl. Acad. Sci. USA*, **81**, 4785–4789.

Wells, J. A. and Powers, D. E. (1966) *In vivo* formation and stability of engineered disulfide bonds in subtilisin. *J. Biol. Chem.*, **261**, 6564–6570.

Wetzel, R. (1987) Harnessing disulfide bonds using protein engineering. *Trends Biochem. Sci.*, **12**, 478–482.

Wetzel, R., Perry, L. J., Baase, W. A. and Becktel, W. J. (1988) Disulfide bonds and thermal stability T4 lysozyme. *Proc. Natl. Acad. Sci. USA*, **85**, 401–405.

Wider, G., Lee, K. H. and Wuthrich, K. (1982) Sequential resonance assignments in protein ^1H nuclear magnetic resonance spectra: glucagon bound to perdeuterated dodecylphosphocholine micelles. *J. Mol. Biol.*, **155**, 367–388.

Wong, C.-H., Chen, S.-T., Hennen, W. J., Bibbs, J. A., Wang, Y. F., Liu, J. L-C., Pantoliano, M. W., Whitlow, M. and Bryan, M. W. (1990) Enzymes in organic synthesis: use of subtilisin and a highly stable mutant derived from multiple site-specific mutations. *J. Am. Chem. Soc.*, **112**, 945–953.

Wuthrich, K. (1986) *NMR of proteins and nucleic acids.* Wiley, New York.

Wuthrich, K. (1989) The development of nuclear magnetic resonance spectroscopy as a technique for protein structure determination. *Accts. Chem. Res.*, **22**, 36–44.

Wuthrich, K., Wider, G., Wagner, G. and Braun, W. (1982) Sequential resonance assignments as a basis for determination of spatial protein structures by high resolution proton nuclear magnetic resonance. *J. Mol. Biol.*, **155**, 311–319.

Withrich, K., Billeter, M. and Braun, W. (1984) Polypeptide secondary structure determination by nuclear magnetic resonance observation of short proton–proton distances. *J. Mol. Biol.*, **180**, 715–740.

Wyckoff, H. W., Hirs, C. H. W. and Timasheff, S. N (eds) (1985) Diffraction mehtods for biological macromolecules. *Methods Enzymol.*, **114**, 3–564.

Yutani, K., Ogasawara, K., Sugino, Y. and Matsushiro, A. (1977) Effect of a single amino acid substitution on stability of conformation of a protein. *Nature (London)*, **267**, 274–275.

Yutani, K., Ogasawara, K., Tsujuta, T. and Sugino, Y. (1987) Dependence of conformational stability on hydrophobicity of the amino acid residue in a series of variant proteins substituted at a unique position of tryptophan synthase α subunit. *Proc. Natl. Acad. Sci. USA*, **84**, 4441–4444.

Zaks, A. and Klibanov. A. M. (1984) Enzymic catalysis in organic media at 100°C. *Science*, **224**, 1249–1251.

Zaks, A. and Klibanov, A. M. (1988a) Enzymatic catalysis in nonaqueous solvents. *J. Biol. Chem.*, **263**, 3194–3201.

Zaks, A. and Klibanov, A. M. (1988b) The effect of water on enzyme action in organic media. *J. Biol. Chem.*, **263**, 8017–8021.

Zale, S. E. and Klibanov, A. M. (1986) Why does ribonuclease irreversibly inactivate at high temperatures? *Biochemistry*, **25**, 5432–5444.

Index

actinidin, 47
activation
  entropy, 131
  enthalpy, 131
  free energy, 131
α-chymotrypsin, 87
α-helix (helices), 31–34
  dipole, 52, 165
  favorite amino acid replacement, 111
  flexibility, 111
  hydrophobic periodicity, 61
  left-handed, 31–32
  protein stability, 111, 157, 159, 160, 179, 183
  protein surface, 61
  right-handed, 31–32
  structural parameters, 31–32
  3_{10} helix, 32; 33
alcohol dehydrogenase, 59
aliphatic amino acids, 48
  aliphatic scale, 107
  hydrophobic interactions, 48, 58
  hydrophobic stabilization, 48, 203
  internal hydrophobicity, 107
  protein interior, 48
  protein stability, 108, 204
amino acids
  buried, 105
  chemical structure, 17
  configuration, 16
  extraction model, 105
  favorite replacement for stability, 111, 166, 179
  fractional accessible area, 105
  hydrophobicity, 105, 106, 163
  mobile, 185
  properties, 18

amino acid side chains
  protein structure, 46–57
amino acid substitution (mutagenesis)
  Ala, 147, 153, 155, 157, 159, 160, 161, 164, 173, 179, 181, 206
  Arg, 147, 149, 150, 166, 167
  Asn, 155, 160, 161, 165, 168, 176, 183, 206
  Asp, 149, 150, 155, 161, 162, 165, 176, 184, 185, 206
  Cys, 144, 161, 173, 206
  Gln, 150, 167
  Glu, 149, 161, 162, 166, 167, 206
  Gly, 155, 157, 159, 160, 161, 164, 179, 181, 206
  His, 147, 181
  Ile, 144, 146, 149, 150, 157, 161, 176, 183
  Leu, 149, 150, 161, 164, 173, 181
  Lys, 157, 467, 184, 185
  Met, 153, 155, 161, 173, 181, 206
  multi-mutation (additional stability), 160, 165, 167, 179, 180, 206, 207, 209
  Phe, 161, 162, 181, 206
  Pro, 155, 156, 157
  protein folded state, 89
  protein folding, 90
  Ser, 147, 149, 150, 151, 153, 160, 161, 165, 168, 173, 181, 206
  stability (see also site-directed mutagenesis), 88, 90, 200
  structural adjustment, 88
  Thr, 146, 147, 149, 150, 151, 153, 157, 167, 168, 176, 181, 184, 185
  Trp, 161, 162, 181, 184, 185
  Tyr, 161, 162, 181, 184, 185
  temperature-sensitive, 88, 184, 185
  tolerant, 88, 185
  Val, 146, 147, 153, 155, 157, 160, 164, 167, 181

Index

aromatic amino acids, 48–50
  hydrophobic interactions, 49–50
  protein interior, 49, 50, 162, 181, 182
  ring-ring orientation, 50
aspartate amino transferase, 88

B. *stearothermophilus*, 84, 102
  glyceraldehyde 3-phosphate dehydrogenase, 112
  neutal protease, 84
  restriction endonuclease, 185
B. *thermoproteolyticus*
  thermolysin, 176
    autoproteolysis, 151
  bacteriophage
  f1 gene protein, 203
  gVp protein, 203
    bacteriorhodopsin, 61
  β-sheets, 34–38
    antiparallel, 34–36
    arch, 36
    β-ribbon, 36, 37
    curl, 36
    hydrophobicity, 34, 35
    parallel, 34–36
    side chains, 61
    twist structure, 34–38
biological modification, 125, 128
bovine pancreatic trypsin inhibitor (BPTI), 83
bulged hairpin, 39

Ca-binding protein, 44
chemical modification, 125, 127
  acetoimidate, 129
  conformational change, 128
  cross-linking, 128
  hydrophilization, 129
  hydrophobization, 129
  key functional residues, 128
  O-methylisourea (guanidination), 130
  salt bridge, 129
  stabilization, 128–130
chemical destabilization (protection), 172
  deamidation of Asn, 172–176
  oxidation of Met, 172, 173
  oxidation of SH, 172
chymotrypsin, 130
collagen, 34
common turn, 38–39
computer model
  configurational entropy, 157, 158
  disulfide bonds, 144–149, 151, 152, 156
  electrostatic interaction, 166
  internal hydrophobicity, 161, 162
conformational entropy, 83, 84
  amino acid substitution, 84, 157
  disulfide bonds, 83
crambin, 53, 205
critical amino acids, 89
cross-linkage
  chemical, 83
  conformational entropy, 83
  disulfide (see disulfide bonding)

Cu, Zn superoxide dismutase, 35, 55
cytochrome c reductase, 49
cytochrome c′, 54

ΔH
  protein folding, 30
  protein unfolding, 81, 82
ΔS
  protein folding, 30
  protein unfolding, 81
ΔH/ΔS compensation, 170, 171
destabilizing substitutions, 89
differential scanning calorimeter (DSC) (see T_m)
dihedral angles
  constraint, 23
  peptide chains, 22
  Ramachandran maps, 23–25
  standard convention, 22
dihydrofolate reductase, 49, 83, 84, 156, 203
dimethylformamide (DMF), 206
dispersion forces
  amino acid substitution, 83
  packing, 83
  protein structures, 27
  protein stability, 83
disulfide bonds
  conformational entropy, 83, 114
  conformational parameters, 56, 144, 150, 154
  engineered dihydrofolate reductase, 84, 156
  engineered subtilisin, 84, 151, 152, 180
  engineered T4 lysozyme, 83, 84, 144, 149
  left-handed spiral, 57, 144, 151, 154, 156
  loop size, 83, 84, 144, 149
  protein stability, 84, 112
  reduction, 148, 155
  right-handed hook, 57, 151, 154
domain, structural, 41, 42

effective concentration, 84
  protein stability, 84
  disulfide bonds (BPTI), 84
electrostatic interactions (salt bridges)
  α-helix, 165
  amino acid substitution, 88, 112
  charged groups, 87, 166
  ionic interactions, 88
  ion pairs, 87
  local polarity, 88
  non-polar side chains, 88
  pH, 88
  protein stability, 87, 110, 112, 165, 166, 206
  protein structures, 27
  unpaired charged groups, 88
entropic effect
  oil drop, 29
  protein folding, 29

ferredoxin, 110, 112
flavodoxin, 35, 36, 65
flexibility (rigidity), 109, 133
free energy change
  determination, 81, 82
  DSC, 81, 160

Index

spectroscopic, 81
pH or denaturants, 82
reversible denaturation, 80–82

glyceraldehyde 3-phosphate dehydrogenase, 112
glycine turn, 38, 39

H-bonds (or H-bonding)
  amino acid substitution, 86
  α-helix (helices), 31
  β-sheets, 31
  free energy, 86
  H-donor, 28
  H-acceptor, 28
  intermolecular, 86
  model compounds, 86
  net work, 168, 169
  protein folding, 86
  protein stability, 86, 87, 110, 113, 167, 206
  temperature-independent ΔH, 86
H-D exchange reaction, 132
  protein structure, 132
  fluctuation, 133
heat capacity (C_p), 81, 90, 201, 208
  protein unfolding, 171
  water accessible surface area, 171
hemoglobin, 43, 112
hybrid enzymes, 131, 132
hydrophobicity (hydrophobic interactions)
  amino acids, 105, 161
  cavity, 85, 163
  ΔC_p, 85
  external, 108
  free energy change, 85
  internal, 107, 161, 163, 164, 179, 181, 182, 203, 204
  packing, 30
  pH-independent, 161, 162
  protein stability, 85, 104–108, 112, 113, 161–164
  protein structure, 29
  scales of amino acids, 106
  temperature-dependent ΔH, 85
  temperature-dependent ΔS, 85
  transfer energy of amino acids, 85
  total, 105
  water-accessible area, 58
hydrophobic amino acids
  distribution probability, 60
  hydrophobic scale, 60
  packing energy, 163
  side chains, 60
  solvent effects, 60
  subunit interaction, 60
  transfer experiments, 60, 163
  water accessible surface area, 163

immobilization, 125–127
  biotechnology, 126
internal environments, 107
inverse turn, 39, 40
ion pairs, 83
iso-1-cytochrome c, 182, 209

kanamycin nucleotidyltransferase, 182, 184, 185
kinetic stability (irreversible denaturation), 83, 131, 132
  aggregation of unfolded state, 147
  chemically damaged unfolded state, 147
  compact structure of unfolded state, 147
  modified proteins
  chemical modification, 131, 132, 202
  chemical protection, 172
  disulfide bonds, 146, 152, 155, 202
  electrophoretic interactions, 166

lactate dehydrogenase, 37, 59, 109, 129, 130, 132, 202
λ phage
  repressor, 84, 160
  cI repressor, 88, 89
  cro repressor, 88, 89
lipase, 205
lysozyme
  T4 phage, 83, 84, 85, 88, 98, 113, 114, 144, 148, 157, 159, 161, 162, 164, 165, 167, 168, 184, 185, 201, 203
  hen egg-white, 174

malate dehydrogenase, 60
maximal stability, 201
model proteins, for mutagenesis
  thermolysin, 176
  subtilisin, 180
mutagenesis
  biological (see amino acid substitution)
  chemical, 184
  thermal, 184, 185
myoglobin, 88
myohemerythrin, 33

non-aqueous solvents
  protein stability, 204, 205
  protein stabilization, 204–207
non-polar side chains (see aliphatic and aromatic amino acids)
  subunit–subunit interface, 58
non-repetitive structures
  bulged hairpin, 39
  common turn, 38–39
  glycine turn, 38–39
  inverse turn, 39–40
  loop, 39–40
  reverse turn, 39–40
nuclear magnetic resonance (NMR)
  protein structure, 199, 200
  protein unfolding, 199
  oligomeric proteins, 57
  symmetry, 58–60
    packing
  cavity, 109, 164, 165
  density, 31
  energy, 203, 204
  favorite amino acid replacement, 109, 111
  hydrophobic interactions, 30, 203
  local packing, 30
  stability, 109, 132, 165

Index

pancreatic trypsin inhibitor, 38
peptide bonds
  dihedral angles, 22
prealbumin, 40, 45
prediction of protein structures, 62–66
prediction of secondary structures, 62–64
  Chou–Fasman method, 62–64
  conformational parameters, 63
  rule, 63, 64
  staphylococcal nuclease, 64
prediction of tertiary structures, 65, 66
  combinational approach, 65
protease
  Strep. griseus, 54
  proteinase K, 154
protein design, 143
protein destabilizing factors, 143
  α-helix formation, 160
  disulfide bonds, 143, 157
  entropy of protein unfolded state, 143, 157
  water, 205
protein stabilizing factors, 89
protein stabilizing forces, 82–90
  conformational entropy, 83, 84
  dispersion forces, 83
  electrostatic interactions, 87, 88, 165
  H-bonds, 86, 87
  hydrophobic interactions, 85
  loop size of disulfide bonds, 84
  role of water, 171, 208
protein structures
  dispersion forces, 27
  electrostatic interactions, 27
  H-bonds, 28
  prediction, 62–66
  van der Waals potential, 27
  multiforms, 79
protein folding
  cooperative, 90
  entropic effect, 30
  ΔH, 30
  ΔS, 30
  interactions, 90
protein homology
  protease, 177, 180
protein interior, 48–50
protein synthesis
  scheme, 19
  transcription, 19
  translation, 19
  codon, 19
  components, 20
protein unfolding
  free energy change, 29, 81
  hydrophobic interactions, 29
  key interactions, 197
  protein stability, 79, 80
  thermodynamics, 29, 80–82

quaternary structure, 57–60
  contacting forces, 57
  oligomeric proteins, 57
  symmetry of aggregates, 57

Ramachandran maps, 23
  non-repetitive structure, 26
  repetitive structure, 24
  twisted β-sheet, 37
recombinant DNA technique
  amino acid sequence, 143
  site-directed mutagenesis, 141–143
repetitive structure, 31–38
reverse turn, 39, 40
ribonuclease, 83, 88, 163, 201, 203
ribosomal protein, 46
rigidity (flexibility)
  B-values, 109
  protein stability, 109
  thermal fluctuation, 109

secondary structure
  repetitive structures, 31–38
  non-repetitive structures, 38–40
  packing, 66
  prediction, 62–64
site-directed mutagenesis (protein engineering), 141–143
  protein stability, 82–89
  protein stabilization (see amino acid substitution)
solvent effects, 61
  hydrophobic side chains, 61
  subunit–subunit interaction, 61
stable proteins
  commercial use, 101
  thermophiles, 101
  mechanism of stability, 103–114
Staphylococcal nuclease, 170, 185, 201
structural type
  all α-type, 42, 43
  antiparallel β, 45, 46
  parallel, α/β, 43, 44
subtilisin, 56, 84, 151, 152, 166, 167, 172, 180, 205, 206, 207
subunit–subunit interactions, 130, 132, 209
supersecondary structure, 40, 41
  βζβ-unit, 41
  coiled–coil α-helix, 40
symmetry of oligomers, 58–60
  point symmetry, 59
  P-, Q-, R-axes, 60, 132

T. thermophilus, 102
T. aquaticus, 102, 180
temperature
  dependent ΔH, 85
  dependent ΔS, 85
  independent ΔS, 85
  independent ΔH, 86
  sensitive, 88, 184, 185
tertiary structure, 41–46
  prediction, 65, 66
thermodynamics for protein unfolding
  reversible denaturation, 80, 81
  ΔG, 81
  ΔH, 81

ΔS, 81
ΔC_p, 81
thermodynamic stability (reversible denaturation), 202
  configurational entropy, 157, 181
  $\Delta H/\Delta S$ compensation, 170, 171
  proteins, with engineered disulfide bonds, 147, 152, 156
  proteins, with engineered electrostatic interactions, 167
  proteins, with engineered internal hydrophobicity, 161–164, 182, 183
  proteins, with engineered H-bonds, 167, 168
thermophiles (thermophilic bacteria), 102
  B. stearothermophilus, 84, 102, 179, 185
  B. thermoproteolyticus, 176
  Pyrodicitum, 102
  T. aquaticus, 102, 180
  T. thermophilus, 102
thermophilic proteins, 103, 104
  aqualysin, 180
  few amino acid substitutions, 111–114
  many amino acid substitutions, 104–111
  neutal protease, 179
  restriction endonuclease, 185
  stability, 103–114
  thermolysin, 178
T_m (melting temperature), 81
  absorbance, 183
  CD, 165, 167
  DSC, 81, 152, 167, 183, 201
tobacco mosaic virus coat protein, 65
transcription (see protein synthesis)

translation (see protein synthesis)
triosephosphate isomerase, 44, 174
trypsin, 130
trypsin inhibitor, 38
tryptophan synthase α-subunit, 85, 87, 88, 113, 181, 202, 203

water
  $\Delta H/\Delta S$ compensation, 170
  disordered structure, 61
  hydrophobic interactions, 60
  ordered structure, 61
  protein bound, 204
  protein structure, 58, 60
  solvent effect, 60
water-accessible surface area, 58
  hydrophobicity, 58, 163
  hydrophobic stabilization, 89
  packing energy, 163
  protein stability, 163
  subunit aggregation, 58

X-ray crystal structure (three-dimensional structure), 143
  crystallization, 143, 199
  dihydrofolate reductase, 156
  λ repressor, 159
  protein engineering, 143
  subtilisin, 151, 152, 172
  T4 lysozyme, 145, 166
  thermal factors, 185
  thermolysin, 178
  triosephosphate isomerase, 175